Digital Transformation and Innovation in Tourism Events

The pandemic has accelerated the digital transformation in tourism and there has been a surge in new, innovative digital initiatives to help tourism businesses. This book provides a comprehensive treatment of the nature of tourism, events and practices in the digital context.

The book looks at how technology has transformed tourism in destination branding, marketing, content marketing, sustainable tourism development and tourism events. It examines the impact of digital transformation on emotions, experiences, information technology tools and marketing techniques.

The book will be a useful reference to those researching tourism, culture, hospitality and marketing and as well as destination planners, managers of tourism destination marketing organizations, regulators, standards and certification bodies, local tourism board authorities and policy makers.

Azizul Hassan is a member of the Tourism Consultants Network of the UK Tourism Society, and has been working for the tourism industry as a consultant, academic and researcher for over two decades. His research interest areas are technology-supported marketing and immersive technology applications in the tourism and hospitality industry, and technology-influenced marketing suggestions for sustainable tourism and hospitality industry in developing countries.

Routledge Advances in Management and Business Studies

Stakeholder Management and Social Responsibility
Concepts, Approaches and Tools in the Covid Context
Ovidiu Nicolescu and Ciprian Nicolescu

Japanese Business Operations in an Uncertain World
Edited by Anshuman Khare, Nobutaka Odake and Hiroki Ishiruka

Entrepreneurship and Culture
The New Social Paradigm
Alf H. Walle

Hospitality and Tourism Education in China
Development, Issues, and Challenges
Edited by Jigang Bao and Songshan (Sam) Huang

Halal Logistics and Supply Chain Management
Recent Trends and Issues
Edited by Nor Aida Abdul Rahman, Azizul Hassan and Zawiah Abdul Majid

Sustainable International Business Models in a Digitally Transforming World
Edited by Anshuman Khare, Arto Ojala and William W. Baber

Cross-Cultural Challenges of Managing 'One Belt One Road' Projects
The Experience of the China-Pakistan Economic Corridor
Arshia Mukhtar, Ying Zhu, You-il Lee, Mary Bambacas and S. Tamer Cavusgil

Digital Transformation and Innovation in Tourism Events
Edited by Azizul Hassan

For more information about this series, please visit: www.routledge.com/Routledge-Advances-in-Management-and-Business-Studies/book-series/SE0305

Digital Transformation and Innovation in Tourism Events

Edited by
Azizul Hassan

Routledge
Taylor & Francis Group

LONDON AND NEW YORK

First published 2022
by Routledge
4 Park Square, Milton Park, Abingdon, Oxon OX14 4RN

and by Routledge
605 Third Avenue, New York, NY 10158

Routledge is an imprint of the Taylor & Francis Group, an informa business

British Library Cataloguing-in-Publication Data
A catalogue record for this book is available from the British Library

Library of Congress Cataloging-in-Publication Data
A catalog record has been requested for this book

ISBN: 978-1-03-222096-3 (hbk)
ISBN: 978-1-03-222097-0 (pbk)
ISBN: 978-1-00-327114-7 (ebk)

DOI: 10.4324/9781003271147

Typeset in Galliard
by Newgen Publishing UK

Contents

Figures

Tables

Contributors

Sharifah Nurafizah Syed Annuar obtained her Doctoral degree in Marketing at the Universiti Malaysia Sabah in 2017. Before that, she studied MSc in Marketing Communications in Middlesex University, United Kingdom. She has passions in the field of marketing communications, social and health marketing, consumer behaviour, tourism marketing and entrepreneurship. She is a Senior Lecturer in Universiti Teknologi MARA Sabah Branch, Malaysia and has served the university for 12 years. She has broad experience in facilitating engagements between governmental bodies and corporations across Southeast Asia and has worked with the community which provides consulting advice to local start-ups and SMEs in Sabah. She is currently appointed as Senior Associate and Consultant of an integrated marketing and business consultancy firm based in Brunei Darussalam. She is also a core member of Southeast Asia Research Academy (SEARA).

Debora Calomino is a journalist, specializing in tourism and territorial marketing. She graduated in Tourism Sciences and Tourism Systems Design and Management at the University of Calabria (Italy) and collaborates with tourism magazines, where she writes about success stories in local marketing.

Beatriz Casais is Assistant Professor of Marketing and Strategy at University of Minho and PhD in Management and Business Studies by University of Porto. She has published in the *Journal of Hospitality and Tourism Technology*, *Journal of Hospitality and Tourism Management*, *Journal of Social Marketing*, *International Journal of Entrepreneurial Behavior & Research*, *Place Branding and Public Diplomacy*, *Review of International Business and Strategy*, *Health Marketing Quarterly*, *Journal of Macromarketing*, *Advances in Advertising Research*, *Tourism & Development*; and *World Review of Entrepreneurship, Management and Sustainable Development*, among others.

Farai Chigora holds a Doctorate in Business Administration (DBA) from University of KwaZulu-Natal (South Africa), a Senior Lecturer in Business Science in the College of Business, Peace Leadership and Governance, Africa University in Zimbabwe. He is a branding specialist with interest in destination branding, strategic marketing, business research and related business areas which he has authored in various refereed international journals.

Tinashe Chuchu holds a Doctorate in Marketing from the University of the Witwatersrand. Currently, he works as a Senior Lecturer in the Marketing Division of the School of Business Sciences at the University of the Witwatersrand, South Africa. Previously, he worked as a Senior Lecturer in the Department of Marketing Management, University of Pretoria, South Africa. He supervises both Doctoral and Masters students in marketing management. Dr Chuchu has published and reviewed for major publishing outlets, which include Wiley, Taylor and Francis, Elsevier, Emerald Publishing, SAGE and presented at the premier conference for marketing academics (the American Marketing Association Conference) which was held in Chicago, USA in 2019. He is a member of the Academy of Business and Retail Management Conferences based in the United Kingdom. He sits on the editorial board of the *Retail and Marketing Review* as well as the *African Journal of Business and Economic Research.*

Katalin Csobán is Senior Lecturer at the Department of Tourism and Hospitality Management at the University of Debrecen, Hungary. She graduated in Tourism and Hotel Management from the College of Commerce, Hospitality and Tourism in Budapest, and earned her PhD in Business and Management from the University of Debrecen in 2010. She has been awarded prestigious research grants to conduct research at Indiana University, USA, as well as at Oxford Brookes University, UK. Her work has been published in renowned international and national journals and books. Her research interests include sustainable tourism development, sports tourism and rural development.

Cynthia Robert Dawayan has been with Universiti Teknologi MARA, Sabah Branch, for the past 14 years where she serves as a Senior Lecturer in the Faculty of Business and Management. She has recently completed her doctoral degree at Universiti Malaysia Sabah, where her research specifically focuses on homestay marketing. Previously attached to the tourism industry, Dawayan's passion and love for the industry and its people continues even after she entered academia. Her research interests are primarily on tourism marketing, entrepreneurship as well as community-based tourism topics.

André Luiz Lopes de Faria received an Undergraduate degree in Geography (Bachelor's Degree) from the Federal University of Juiz de Fora in 1993, Social Studies from the Higher Education Center of Juiz de Fora in 1992, Master's in Environmental and Forestry Sciences from the Federal Rural University of Rio de January in 2001, and a PhD in Agronomy (Soils and Plant Nutrition) from the Federal University of Viçosa in 2010. He is currently an Adjunct Professor at the Federal University of Viçosa (UFV). He coordinates the DGE/UFV Quaternary Geomorphology Laboratory. CHe also coordinates the research group on Continental and coastal landscapes. He is a supervisor of Master's students in the courses of Cultural Heritage, Landscapes and Citizenship, and in Geography at UFV.

Syed Arslan Haider is a PhD scholar in the Department of Management at Sunway University Business School, Malaysia. He received his BS in Computer Science and Master in Project Management from the Capital University of Science and Technology, Pakistan. He is active in research in the areas of Knowledge Management, Innovation, Leadership, Organizational Culture and Project Complexity. His research has appeared in reputable journals such as *Journal of Knowledge Management, Abasyn Journal of Social Sciences* and many others Scopus indexed journals. He is a Project Manager of Soulmate Construction Company in Pakistan.

Imran Hasnat is Assistant Professor of Digital Journalism at the Gaylord College of Journalism and Mass Communication, University of Oklahoma. He is a co-PI on a Bureau of Educational and Cultural Affairs (ECA) grant with South Asia and has participated in the Studies of the US Institutes (SUSI) program in 2011. His research interests are in the use of digital media in public diplomacy, grassroots media such as community radio and the role of tourism advertising in establishing and promoting country image. He teaches multimedia, digital reporting, data and visualization to future journalists. Imran has a PhD and a Master's degree from the University of Oklahoma, as well as a Master's and Bachelor's degree from Jahanginagar University in Bangladesh.

Azizul Hassan is a member of the Tourism Consultants Network of the UK Tourism Society. Dr Hassan has been working for the tourism industry as a consultant, academic and researcher for over two decades. His research interest areas are technology-supported marketing for tourism and hospitality, immersive technology applications in the tourism and hospitality industry, and technology-influenced marketing suggestions for sustainable tourism and hospitality industry in developing countries. Dr Hassan has authored over 150 articles and book chapters in leading tourism outlets. He is also part of the editorial team of 25 book projects from Routledge, Springer, CAB International, and Emerald Group Publishing Limited. He is a regular reviewer of a number of international journals.

H.M. Kamrul Hassan is currently serving as Assistant Professor in the Department of Marketing, University of Chittagong. His principal interest in teaching is developing, using and disseminating comprehensive, affordable teaching and business education platform to be maximized both by faculty members and students. He studied and lectured in sustainable marketing, entrepreneurship and tourism. He has published articles in national and international journals. Moreover, he has attended many local and international conferences in Bangladesh, India, Malaysia, Thailand, Indonesia, Czech Republic and Germany. In addition, he regularly writes on concurrent business issues in academic and non-academic journals and newspapers.

Md Sazzad Hossain is Research Associate at School of Hospitality, Tourism and Events, Taylor's University, Malaysia. Recently, he was awarded a PhD in Hospitality and Tourism from Taylor's University, Malaysia. His research interest is consumer behaviour, competitive advantage and tourism sustainability.

Md. Wasiul Islam is Professor in Forestry and Wood Technology Discipline of Khulna University, Bangladesh. He holds a PhD in Tourism from Business School of the University of Queensland, Australia. Before his PhD he completed two Master's programs. He did his first MSc and BSc (Hons) from Forestry and Wood Technology Discipline, Khulna University. He did his second MSc in Forest and Nature Conservation (Minor in Leisure, Tourism and Environment) from the Wageningen University and Research Centre, the Netherlands. His research interests are focused on nature-based tourism (particularly community-based tourism and forest-based ecotourism), shared governance, participatory management of protected areas, co-management approach, sustainability, community development, and blue economy. He has published his research findings in several international peer-reviewed journals.

Muhammad Abdul Kamal is working as Assistant Professor in the Department of Economics at the Abdul Wali Khan University Mardan, Pakistan. He also worked as Postdoctoral Research Fellow at the Henan University, China. His research interests include China's FDI, trade competitiveness and potential, institutional environment environmental quality and tourism. He has published articles in peer-reviewed international journals such as the *Journal of Asia Pacific Economy, Emerging Markets Finance & Trade, Utilities Policy,* and *Economic Research-Ekonomska Istraživanja,* and *Environmental Science and Pollution Research.*

Joice Lavandoski is Adjunct Professor at the Department of Tourism and Heritage at the Federal University of the State of Rio de Janeiro (UNIRIO); an invited Professor in the Graduate Program in Tourism at the Federal Fluminense University (UFF), Brazil; and the Coordinator of the Tourism Events Laboratory (UNIRIO). Lavandoski graduated and attained a Master's Degree in Tourism from University of Caxias do Sul (UCS), Brazil; and a PhD in Tourism from University of Algarve (UALG), Portugal.

Annaelise Fritz Machado had Undergraduate Degrees in Tourism from the School of Tourism of Santos Dumont (2001), Brazil, in People Management (2019), in Business Administration (2020) and in Marketing (2020) from Estácio de Sá University, Brazil. Machado also had Graduate Degree in Organization and Administration of Leisure and Recreation and Events (2004) from the Federal University of Juiz de Fora, Brazil. At present, she is a Master's student in Tourism Management at the Polytechnic Institute of Cávado and Ave, Barcelos campus, Portugal.

Shirzad Mansouri is the founder of Talent Forte Training Center in Bangkok, and a University Lecturer. He obtained his Bachelor in English and Master Degree in teaching English as Foreign Language (TEFL). He also received his MBA in Tourism and Hospitality, and Doctorate in Business Administration.

He has more than 25 years of experience working alongside the executive team of different educational and academic institutions in Iran, Thailand, and Indonesia. Shirzad specializes in leadership and tourism and hospitality management. He has been the head of tourism and hospitality curriculum for undergraduates since 2019. His current research looks at the Human Capital Development during COVID-19 and the survival of tourism and hospitality businesses in Muslim communities in Thailand, Indonesia, and Iran. Pursuing a tourism research agenda, he has recently been working on TEFL tourism in Bangkok. His area of interest is technology and tourism development, tourism entrepreneurship and innovative education in human capital development in tourism.

Brighton Nyagadza is Full-time Lecturer and A/Chairperson of the Department of Marketing (Digital Marketing) at Marondera University of Agricultural Sciences and Technology (MUAST), Zimbabwe, Full Member of the Marketers Association of Zimbabwe (MAZ), an Associate of the Chartered Institute of Marketing (CIM), United Kingdom and Power Member of the Digital Marketing Institute (DMI), Dublin, Ireland. His research expertise revolves on corporate storytelling for branding, digital marketing, Public Relations (PR), marketing metrics, financial services marketing, and educational marketing. He has published several book chapters in Routledge Books of Taylor and Francis Publishers, New York (USA), Lexington Books of the Rowan and Littlefield Publishers, Maryland (USA), Langaa Publishers (Cameroon) and in reputable global journals such as those published by Intellect, Bristol, UK; SAGE, London, UK; Taylor and Francis, UK; University of Johannesburg, South Africa; Westburn Publishers, Scotland; UNISA Press and others.

Mohammed Shahedul Quader studied and lectured International Business, Marketing, and Shipping Management, working as Associate Professor in the Department of Marketing, University of Chittagong. He has completed his MBA in Shipping Management, MA in Marketing Management from University of Leicester and Middlesex University, UK respectively. He has more than 15 years of teaching experience in the field of international business. Besides teaching, he has published more than 30 articles in reputed local and international journals and joined international conferences locally and abroad.

Muhammad Khalilur Rahman currently works in the Faculty of Entrepreneurship and Business at Universiti Malaysia Kelantan (UMK), Malaysia. He has been actively involved in research activities and has published over 55 articles and book chapters in leading service management outlets. He has a wide interest in tourism and service management research which includes medical tourism, eco-tourism, halal tourism, service quality, brand equity, supply chain management, operation management, green and sustainable development. Rahman is a reviewer of *Kybernetes, Journal of Hospitality and Tourism Insights, International Journal of Contemporary Hospitality Management, Journal of Islamic Marketing*, and *BMC Public Health*.

Shafiqur Rahman's qualifications include PhD, MBA, BA and Marine Engineering. He teaches Information Systems, Marketing and Management courses at the undergraduate and postgraduate level in Sydney, Australia. He is a visiting professor for United International University, Bangladesh. He has presented in 30 international conferences, published 20 research papers (including ABDC, ERA and SCOPUS ranked journals), four book chapters and one book. He supervises PhD students, examines PhD theses, reviews journal articles and mentors scholarly activities. He served in the industry for 15 years and has been in academia over a decade.

Samik Ray is an ex-faculty of Department of Folklore, University of Kalyani; Travel and Tourism Management in MPTI (Kolkata); and WTCC School of Trade and Commerce (Kolkata); ex-trainer and Faculty of Regional Level Guide (RLG) Training (Department of Tourism, Government of India), presently working as RLG (Department of Tourism, Government of India); Visiting Faculty in the Department of Tourism Management, Rāmākrishna Mission Vidyāmandira Autonomous College with potentials for excellence, UGC; Editor of "Tourism Theory and Practice", the author of several essays on tourism studies and management, social science, and literary criticism, and received the National Tourism Award in the category of "The Best Tourist Guide".

Nuria Recuero-Virto is currently employed as Assistant Professor at Universidad Complutense de Madrid. She is now in the Deanship of the Faculty of Commerce and Tourism, as Delegate for the Dean for Institutional Communication and Digital Transformation. She was awarded a Post-Doctoral (2014–2018) and Pre-Doctoral Scholarship (2010–2014). Due to this background, her specific areas of interest are: tourism marketing, employer branding and neuromarketing. She was finalist of FITUR's awards for best doctoral thesis (2013). Her research has been published in journals such as *Journal of Destination Marketing & Management; Journal of Hospitality and Tourism Management; Tourism Review,* among others.

Anum Rehman is a PhD aspirant at the Busines School of Sunway University, Malaysia. She has completed her BSc honours in Marketing and sales and MBA General from Forman Christian College (A Chartered University), Lahore. She has been working for the past nine years within the IT industry and edu-cation industry. She aspires to learn and apply her knowledge for marketing and practice. She is currently Program Manager at Lahore Institute of Future Education, Lahore.

Gabriela Oliveira Rodrigues graduated in Agroindustry Technician by the Federal Institute of Southeast of Minas Gerais, 2016. Rodrigues also has a degree in Tourism Management from the Federal Institute of Southeast of Minas Gerais, and was a scholarship holder of the Scientific Initiation project "Touristic Remembrance: Influencing Consumption Aspects of the Federal Institute of Southeast Minas Gerais". Rodrigues participated in the IF Sudeste

MG International Mobility program, conducting a survey over six weeks in Portugal. Rodrigues is currently a Master's student in the Graduate Program in Landscapes, Cultural Heritage and Citizenship at the Federal University of Viçosa.

Rolee Sifa is currently pursuing a Diploma in Information, Digital Media and Technology from the Business and ICT department of TasTAFE, Tasmania, Australia. She is an enthusiastic and self-motivated student in educational research. Her areas of research interest are digital communications, business, entrepreneurship, leadership, health science, and lifestyle.

Bruno Sousa is Adjunct Professor of Marketing at Polytechnic Institute of Cávado and Ave (IPCA) and PhD in Marketing and Strategy by University of Minho (Portugal). He has published in the *Journal of Enterprising Communities; Enlightening Tourism; Quality – Access to Success; Revista Brasiliera de Gestao e Desenvolvimento Regional; European Journal of Applied Business and Management; World Review of Entrepreneurship; Management and Sustainable Development; Turismo y Responsabilidad Social; International Journal of Public Sector Performance Management; Revista Iberica de Sistemas e Tecnologias de Informacao; Smart Innovation, Systems and Technologies; Dos Algarves* (a multidisciplinary e-journal); *International Journal of Marketing;* and *Communication and New Media*, among others.

Elanie Steyn is Associate Dean and Associate Professor at the Gaylord College, University of Oklahoma, USA. She teaches and researches in management, leadership and business. She directs the US Department of State/University of Oklahoma grants that involve South Asian students, entrepreneurs and media professionals. She is an editor of *Global Journalism Education in the 21st Century: Challenges and Innovations* (Knight Center for Journalism in the Americas) and co-editor of *Critical Perspectives on Journalists' Beliefs and Actions. Global Experiences* (Routledge). She earned an MA in Business Communication from Potchefstroom University, South Africa, an MA in Communication Policy Studies from City University, London and a PhD in Business Management at North-West University, South Africa.

Shehnaz Tehseen is a Senior Lecturer of Entrepreneurship and Program Leader of BSc (Hons) International Business at the Department of Management in the Sunway University Business School, Sunway University, Malaysia. She received her Master of Business Management and PhD (Management) from International Islamic University Malaysia (IIUM) and Universiti Kuala Lumpur Business School, UNIKL, Malaysia respectively. She received Best Student Award in MBA (General Management Specialization) in 27th convocation n October 2011. She also received Best Paper Award for conference paper entitled "Ecological Perspective of Firm Innovation: Implications for Entrepreneurship Success" in the ICBSI 2018, 17–19 October 2018. Her research areas include Entrepreneurship; Ethnic Entrepreneurship; Innovation; Wholesale and Retail SMEs; Knowledge Management; Human Resource

Management; Marketing and Consumer Behavior; Sustainable Development; and Smart Cities. Her articles have appeared in *Business Process Management Journal, Asia-Pacific Journal of Business Administration, Journal of Research in Interactive Marketing, Spanish Journal of Marketing-ESIC, Production, International Journal of Entrepreneurship* and many other Scopus indexed journals. She is an Assistant Editor for *Entrepreneurship of Journal of Global Business Insights* and International Advisory Board member of *International Journal of Entrepreneurship*. She is also a member of Academy of Management (AOM) and Association of North America Higher Education International (ANAHEI).

Sweta Thakur is Head of Program for Bachelor of IT (BIT) at King's Own Institute (for over three years), Sydney, Australia. In addition to her leadership role for BIT, she has been teaching Information Technology courses since joining. She also held leadership positions in other similar organizations. Prior to coming to Australia, she was an Assistant Professor of Computer Science at the Government College of Engineering and Leather Technology during 2009 to 2015. Sweta received a PhD from University of Javadpur, India. She is a member of IEEE, USA, IEEE Gold Affinity Calcutta and Women in Engineering.

Hasanuzzaman Tushar is Assistant Professor of College of Business Administration at the IUBAT-International University of Business Agriculture and Technology, Bangladesh. He is a PhD candidate at the Graduate School of Human Resources and Organizational Development, National Institute of Development Administration (NIDA). He also served at Chandigarh University as a visiting professor. His research interest primarily lies in the area of Career Development, Human Resource Development and Management, Tourism Education, Leadership, and Social Stratification. He authored various publications in well-indexed journals and presented several papers at national and international conferences.

Kaplan Uğurlu is Associate Professor at Faculty of Tourism in Kırklareli, Kırklareli University, Turkey. He has been working in Turkey for 11 years as an academician. He has worked as a senior manager for 25 years in the tourism sector. He is specialized in marketing, finance, accounting, cost controlling and hotel openings. After his bachelor's degree at Uludağ University, Turkey (BSc in Tourism and Hotel Management), he completed his master's degree at University of Surrey, England (MSc in International Hotel Management) and received his PhD at Marmara University, Turkey (PhD in Production Management and Marketing). He received his associate professor degree in 2019 by the Interuniversity Board Presidency of Turkey. He has more than 55 papers presented and published in national and international congresses, journals, and books. His academic research interests include tourism and hotel management, tourism and hotel marketing, accounting, and finance.

Yuanyuan Zong, CHE, CHI & CHIA (AHLEI) is a PhD candidate from Graduate Institute of Marketing and Tourism Management, National Chiayi University, Taiwan. She is also a Hospitality Lecturer at Wuhan Business University, China, and a Visiting Scholar at Oklahoma State University, USA. Her research interests are cultural tourism, destination marketing, hotel operation management and hospitality and tourism education. She has published more than 30 journal papers.

Introduction

Azizul Hassan

Digital innovation in tourism events is an important motivation of better planning, promotion and marketing. The influences of technology in tourism events are increasing across the world. Digital innovation has been widely implemented in the tourism events to minimize costs, increase operating performance, and, most significantly, improve service quality and experience. Digital innovation also aids in the evaluation of alternative events as well as making comparisons and contrasts of available options. The main goal of digital innovation, which brings together tools to facilitate growth, usage and knowledge sharing, is to make tasks easier and to solve many of the problems of tourism events.

It was extremely difficult for the tourism industry to showcase events to consumers prior to the advent of digital innovations. It was also very costly because consumers are often physically separated. However, the introduction of digital innovation has simplified business transactions while still increasing the customer base. Inter-organizational systems, which linked organizations, were once the most common type of technology. However, several companies were unable to afford to use them due to the high costs. The advent of global distribution networks made for fast cross-border links and communication. This allowed for quick knowledge transfer, which helped the industry grow in terms of bookings. Global distribution was an inter-organizational structure that arose from computer reservation systems that assisted in the integration of airline information. Customers were able to make their reservations at one traditional marketplace, which boosted the tourism events industry. These devices were in use in the 1960s, and it was difficult to integrate them with modern computers. The Internet and the World Wide Web (WWW) were invented as a result of digital innovations, and they have changed the way people communicate and conduct business. Businesses use the Internet to promote events. This forum has brought the whole world together. It has transformed it into a global community. People can communicate in real time in different locations. They can exchange ideas and thoughts without having to meet in person. Digital innovation has resulted in the development of various types or networks that allow for interconnection.

Tourism events have been transformed as a result of how digital innovation has changed their ways of doing things. Digital innovation has an effect on how tourists interact, read and think about an event. It contributes to culture and

DOI: 10.4324/9781003271147-1

influences how they communicate on a regular basis. Tourists live in a time when digital innovations are commonplace. Mobile phones and the Internet are two examples. Thus, the tourism event industry is one of the industries that has seen the most improvements as a result of digital innovations. It is unsurprising that this sector is regarded as one of the most profitable in the world. Digital innovation has played a significant role in this achievement. Despite some shortcomings as unequal availability, digital innovation is critical to the growth of the tourism events industry. At both regional/domestic and international levels, the tourism event industry is one of the most successful. The adoption of digital innovation in its management and operations is the driving force behind this rapid development. Many tourism events companies integrate digital innovation into their operations, and as a result, stand a strong chance of reaping significant benefits. Computerized reservation systems are one type of digital innovation system that is used in this industry to reach out to potential customers. However, digital innovation has both positive and negative effects on tourism events, as well as a direct influence on tourists' lives coupled with a crisis such as COVID-19.

This book is thus uniquely designed with very focused contents (i.e., conceptual discussions, cases and future research directions) followed by critical explanations, examples and in-field author accounts. Any specific attention has not been laid on conventional theories and models on technology adoption/acceptance/diffusion that are commonly rigid statistical analysis reliant to examine the adoption of technologies within its limited space. Also, this book carefully bypasses proposing any business models to investigate how technologies help event organizers to disrupt, innovate or change their business model, and so on. Still, this book looks at the theoretical contexts and how technology affects and is used on both the supply and demand sides of tourism events. The contents of this book cover tourism events having business, community, religious, sports, and other features. The book analyses digital innovation use in tourism events in order to present the facts and the function of marketing techniques in tourism event promotion. A short summary of the background and objectives of all chapters is presented below.

In Chapter 1, Quader and Hassan deliberate theoretical foundations and a conceptual debate of tourism events based on digital innovations. This research looks at how tourism and technology interact, with a particular focus on how digital innovation is integrated into tourist events. In this regard, this research used a comprehensive literature review to determine the roots of digital innovation-based tourism events and to develop a conceptual framework that explains how the integration of digitalization may provide value for all system stakeholders.

In Chapter 2, Islam analyses the popularity of Bangladesh's natural-based tourist destinations. Tourists usually visit these locations for leisure and recreation. There are only a few notable activities that take place specifically at these locations. There is no scientific literature that covers certain events that occur in these nature-based destinations from the perspective of Bangladesh. As a result, the aim of this research is to look into events held at Bangladesh's nature-based destinations, as well as the role of information, communication and technology (ICT) in their marketing.

In Chapter 3, Zong and Kamal explore light show technology (LST) as an effective tool for drawing visitors to traditional festivals and city centres. The impact of LST on the formation of tourism destination image and the effect of tourist affective cognition in the sense of city and traditional festivals has received little attention in the past. This research aims to review relevant literature and summarizes landscaping cases of light show technology performed by major Chinese suppliers, in order to provide a current status introduction on light show implementation in the tourism festival industry. Then, by experiencing the lightscape in China's major cities and traditional festivals, a mixed study of visual images and qualitative interviews are used to investigate tourism elements and tourist affective cognition.

In Chapter 4, Ray argues that the coevolution of a digital application supported network of festival-event focused tourism stakeholders and digitally armed consumers resulted in the emergence of a new ecosystem. The scope of digital application in Kolkata's Durga festival, an important tourist motivator for the city, is examined in this chapter.

Annuar and Dawayan focus in Chapter 5 on the importance of Harvest Festival, also known as Kaamatan, the highlight of all activities in Sabah, Malaysia. Sabah, located in the northern part of Borneo Island, is home to more than 30 ethnic groups, each with their own culture, customs, values and language. The festival is a cultural event that takes place every year on 30 and 31 May to thank the gods for a bumper harvest.

In Chapter 6, Haider, Rehman and Tehseen observe that organizing activities on a digital platform seems to be a difficult challenge. The aim of this analysis is to see how ICT factors affect major events like Coke Fest 2020 and the Pakistan Super League (PSL) in the years 2019 and 2020. The Coca-Cola Food and Music Festival is a music-food mashup that brings Pakistanis' two passions together in one venue.

Mansouri in Chapter 7 analyses the case study of a local food festival in Thailand that uses technology in tourism events. Thailand is developing as a gastronomy tourism destination, with everything from street food to foreign cuisine available. The practical implications of business intelligence (BI) and business analytics (BA) in the Thai tourist sector are investigated and explored in this study. A case study from a local Thai cuisine festival was used to analyse these.

In Chapter 8, Uğurlu evaluates innovative technology efforts to improve the attractiveness of Turkey's event and festival tourism. Turkey is one of the world's most popular tourist destinations and is also one of the most forward-thinking countries when it comes to event tourism. It is strange that in Turkey, where over a 1000 national festivals and events are conducted each year, the number of international festivals is so low. The usefulness of technology at events and festivals, as well as its advantages to Turkish tourism, are all carefully investigated in this study.

Nyagadza, Chuchu and Chigora in Chapter 9 reconnoitre technology application in tourism events in Africa. The goal of this chapter is to look at how technology has impacted the tourist business in Southern African countries. This is accomplished by evaluating the important digital tourism characteristics related with each of the states addressed in a methodical manner.

In Chapter 10, Tushar, Rahman, Thakur and Hossain investigate the ubiquitous role of mobile technology application in the Australian Open. This study uses the content analysis method to look at how technology is influencing events in Australia. The country is one of the most popular sports tourism destinations on the planet. The tourism industry employs a large portion of Australia's workforce and makes a significant contribution to the national economy.

In Chapter 11, Csobán briefs about technological innovations in the 2021 Sabre World Cup in Budapest in Hungary. The goal of this study is to look at how technology is used in the pre, during, and post-travel stages of sport tourism. The latest technical innovations in athletic events are explored, with a specific focus on the Sabre World Cup, Budapest, due to be held in Hungary in 2021. The current study included a qualitative approach, which included on-site observations, a content analysis of the webpages, and in-depth interviews with event organizers.

In Chapter 12, Calomino outlines the applications of technology in Note di Fuoco Festival in Calabria in South Italy. Every year in Calabria, a city in southern Italy, a festival dedicated to pyrotechnic art called Note di Fuoco takes place. This chapter looks at the technologies that are used to put on this event, which happens every year in July, both in terms of choreography with fireworks and music, as well as in terms of media promotion and experience facilitation (online ticket purchase, booking, post-event experience, tourist information in the area, applications for enjoying the live event).

Chapter 13 is written by Sousa and Casais and discusses the "7 Gastronomical Wonders" event and communication digitalization in the Portuguese context. A survey was carried out to measure the attitude of respondents to experience, not only in the degustation but also in agriculture and vineyards, or in cultural activities in those places associated with food and wine. In order to better comprehend the study phenomena a social media analytics will be built. In its multidimensional structure, the place connection is analysed.

In Chapter 14, based on the previous debate, Recuero-Virto sheds light on the importance of technology efforts as a driver of the event tourism sector's revival in Spain. This chapter's interest stems from the undeniable requirement for technological change in the tourist business in the post-COVID-19 environment.

In Chapter 15, Steyn and Hasnat focus on the period following January 2020 that has posed (and continues to pose) unparalleled problems to the planet. This chapter will discuss how technology, applications and big data are helping the tourism industry get back on its feet after the Great Recession, from livestreaming to updated 5G technology, facial recognition and AI to scannable QR codes that ensure touch-free ordering in restaurants, mobile room keys, and cloud-based software to coordinate staffing duties and help operators comply with newly enhanced cleaning standards.

Machado, Sousa and Lavandoski in Chapter 16 outline technology applications in Rock in Rio Brazil as a tourism event from a stakeholder perspective. This chapter seeks to examine the innovations made by the event-sponsoring parties in the newest edition of Rock in Rio Brazil, how the technology influences the event, generates entertainment and the visitor movement. While the event is cemented in the country schedule of events, scientific study on the issue is lacking and this work offers additional uniqueness and worth.

In Chapter 17, Machado, de Faria and Rodrigues feature the sacred in cyberspace of the Taper of Our Lady of Nazareth religious event and technology application in the (re)construction of territorial and touristic identities in Belém do Pará, Brazil. The purpose of this chapter is to exhibit The Taper of Our Lady of Nazaret and to analyse how online technologies assist in the re(construction) of territorial and tourism identities in Belém do Pará, Brazil. The study object is The Taper of Our Lady of Nazareth, and the approach utilized is descriptive research. What kinds of Internet technologies contribute to the re(construction) of society? What forms of Internet technologies contributing to the re(construction) of territorial and tourism identities in Belém do Pará, the subject of this study? The findings show that the cyberspace event is an innovative and interactive religious experience that allows for interaction through technology-mediated behaviours. This article covers inputs in the disciplines of tourism, technology and events in the context of the world's largest religious event from an interdisciplinary approach.

Rahman, Sifa and Hassan in Chapter 18 explain the effects of COVID-19 on tourism events, technology advancement and future research directions. With COVID-19 outbreaks, the relevance of technological acceleration, and its influence on tourism events, this chapter seeks to examine the rethinking of tourism and tourist events framework. This study employs a research technique to synthesize current literature and concepts in the context of digital transformation in the tourist events sector, as well as future research directions in light of the COVID-19 pandemic's continued uncertainty.

The contents of this book therefore discuss the continuous rapid rise of digital innovation applications in tourism events around the world, which can serve as a powerful and fundamental source of assistance in a variety of ways.

Part I
Conceptual Discussions

1 Digital Innovation in Tourism Events

Theoretical Underpinnings and Conceptual Discussions

Mohammed Shahedul Quader and H.M. Kamrul Hassan

Introduction

Event tourism industry is flourishing all around the globe, and its nature, scope and purpose differ from other types of regular tourism activities (Higgins-Desbiolles, 2018). Events are at the central point of swift socio-cultural and technological change. The World Wide Web has brought a dramatic revolution in the form of integrated information systems that escalated digital mobile-based tourism to help create, communicate and deliver service in time effectively (McCabe et al., 2012). The advent and convergence of smart technology are penetrating and reshaping the event tourism sector through an array of well-established technologies. With the development of sophisticated technological progress, the event industry is moving forward towards an innovative economic avenue by connecting technologies with market needs (Mitchell et al., 2016). Faster Internet facilities and information technology have ushered in a spectacular transformation that will continue to foster networks across many stakeholders and organizations, ultimately leading to digital innovation. Digital innovation is becoming increasingly important across all industries and functional units, and consequently, a growing number of strategic stakeholders employ digital technology to foster innovation (Grossman, 2016). As the world becomes increasingly digital, cultural institutions are paying greater attention to the opportunities afforded by digital innovation, especially in the ever-complex and demanding tourism sector (del Vecchio et al., 2018). Digital innovation may help spur the development of new business models in the tourism industry since it provides tourists with a better experience (Li, 2020).

In the minds of tour participants, an event is a reflection of the arrangement, settings, physical efforts, socialization, fulfilments and psychological engagements of the event. As an area of study, event tourism is described as how festivals and events may bring prospective tourists to a community or geographic area, which helps a destination grow economically and regionally. Event tourism is rooted in the concept that events and festivals are a form of tourism, which aims to bring

DOI: 10.4324/9781003271147-3

out responsible tourism activities while also expanding tourism diversity since it significantly impacts both tourists and communities (Gursoy et al., 2020; Raj and Musgrave, 2009). Globally, tourists nowadays seek a comprehensive, integrated tourism package that aids them in comprehending the essential amenities and support systems offered at event as well as festival location.

Digital innovation is all about the development and implementation of innovative goods and services that challenge, substitute, or supplement current terms of the system within organizations and industries (Hinings et al., 2018). New products, procedures, services, applications and even business models are all part of digital innovation when they are carefully orchestrated within a specific environment (Nambisan et al., 2017). Since digital innovation management promises a rich and highly gratifying study topic for academicians and researchers, innovative conceptualization on the topic is urgently required (Nambisan et al., 2017). Therefore, unconventional thinking on digital innovation management that more appropriately deals with the continually evolving nature of innovation systems in a digital ecosystem is important. Event tourism is experiencing new technological advancements, which necessitates exploring digital innovation to find solutions to previously unaddressed issues. Through digital innovation, event tourism has the potential to reshape the association with different hotel chains, logistics and supply chain firms, online and telecommunication companies, travel agencies and other stakeholders directly engaged with the event tourism sector, which may ultimately have a big influence on tourism development in different communities all over the world.

An integrated service package involves a regular awareness and synthesis of diverse visitor behaviour patterns and expectations, as well as digital technological advancements facilities, reengineering management systems and coordinated stakeholder engagement (McCabe et al., 2012). Therefore, modern tourist business integrates technology with a wide range of events, festivals and restorations of historic sites that encourage visitors through active engagement with the attractions as well as enthusiastic participation in tourism fairs, events and exhibitions. However, there have been relatively few studies conducted in order to evaluate the theoretical underpinnings and conceptual debate on the effects of digital technology on tourist events and festivals, which has prompted the consideration of this study. This qualitative study evaluates the theoretical foundations and conceptual discussion of digital technology on tourist events and festivals, as well as addressing the competitive advantage and digital innovation in tourism events and festivals. This research will highlight events and festivals that fall under the broad term "tourism events". The findings will help to provide a conceptual overview of theories of tourist events and festivals while also laying the basis for future studies. Tourism event and festival organizers can comprehend how digital technology may create a strong influence on tourism events and festivals, which may shape the future tourism economy.

This study used a qualitative method to achieve well-defined research goals and to obtain a comprehensive understanding of the theoretical underpinnings and conceptual discussion of digital technology in tourism events and festivals. This is a desk-based research analysis by definition, with a focus on the interpretivism method of research philosophy (Goldkuhl, 2012).

Digital Innovation Driven Events Tourism

Sophisticated innovative technologies (e.g., Internet, ICT and digital technology) are driving the dominant economic wheel of the event tourism industry in the virtual environment (Breukel and Go, 2009; Li et al., 2018; Murphy et al., 2016; Willis et al., 2017) and reshaping the travel and tourism management style to organize and control the tourism resources in order to deliver value-generating service with economies of scale (Cimbaljević et al., 2021; Xu et al., 2019). Garbelli et al. (2017) postulated that ICT could be applied to offer complete and effective virtual communication and inform tourists about multiple implications of value-generating programs of destinations. While digital technology has traditionally interlinked enterprise to enterprise and industry to industry, Internet technologies and particularly social media create innovative opportunities for both business and stakeholders in the form of prompt service, value delivery and better relationship (Leung et al., 2013; Sigala et al., 2016; Wozniak et al., 2017). This trend will continue to change in a positive direction and have strategic business management implications within the tourism Industry atmosphere (Berne et al., 2012). It has been evident that Internet technology accommodates and deliver a huge amount of required tourism information which may foster digital innovation among tourists and various actors within the industry environment. The study of digital innovation brought drastic changes in strategic business framework and event tourism market structure in developed and emerging countries. With the introduction of digital innovation in the technology era, the systematic strategic analysis and interplay with gigantic information create remarkable market opportunities for the event tourism industry. Digital innovation is characterized by a shift from one range of legitimated, current institutional structures to another set of legitimated, updated institutional frameworks while maintaining a focus on legitimacy (Hinings et al., 2018). Digital technologies create a broad scope for event tourism enterprises' activities, and it is prudently expected that future success and long-term survival of event programs can depend on digital stability and transformation (George and Paul, 2019) which promotes digital innovation.

Diversely updated technologies support tourists at various levels and enhance comfort zones for them, comprising an easy information search, prompt decision making, on-site experience and booking for accommodation and programs (Grissemann and Stokburger-Sauer, 2012; Jin and Phua, 2016; Murphy et al., 2016). Information and communication technology has already become a driving force of the smart tourism system to serve tourists and other stakeholders with huge authentic data, greater mobility and rational decision making (Sigala and Chalkiti, 2014). In the present business atmosphere, enterprises involved in management and marketing event tourism require careful accommodation of contemporary and complementary spatial information. On the other hand, knowledge of tourist destinations represents a vital strategic issue when deciding to tour and travel. A broad range of information and communication technology networks can be applied to maximize the innovation-oriented event experience. The explosive growth of IT forcing even tourism to move forward and exploit new market opportunities in order to collectively satisfy stakeholders (Lee et al., 2016). The event tourism industry has already been transformed by digital proliferation and business process innovation, where IT actively plays a decisive role, boosting tourist experiences on a greater scale event and creating better scope for the management of enterprises (Neuhofer et al., 2013).

Evaluation of Models Related to Digital Innovation and Situational Realities

A number of well-recognized established models have been widely applied in tourism research to evaluate the behavioural effects of tourists from various dimensions (Table 1.1). The consequence and impact of technological change within the virtual environment and event tourism industry require evaluation with further thought. Policymakers need to have a clear understanding of the hypermobility of virtual atmospheres (minimizing temporal and spatial periphery) that create interdependent value-generating systems for achieving diverse socio-economic objectives (van Nuenen, 2016). Reengineering of the tourism industry takes place by cultures of diffusive networks and ubiquity of advanced technology in various forms (Hannam et al., 2014). The emergence of cyberspace redesigns and mobilizes the nature of space itself, where virtual spaces are based on public interest rather than physical vicinity (Hannam et al., 2014). Such types of spatial conversion create new cultures of co-creation in the business system as beneficial exchange relationships between organizations and customers are conflated at a higher degree (Campos et al., 2018).

The tourism industry has been moving forward with remarkable performances of pseudo-independence of smart technology that pervade our social lives and create opportunities to acquire information that helps with real-time rational purchase decision making (Lamsfus et al., 2015). The recent tourism has sprung up as omnipresent socio-technological interdependent relationships escalating socialization processes and connectivity to share information and knowledge for making realistic decisions (van Nuenen, 2019). The researchers have observed that tourism enterprises seek to explore techno-social opportunity by virtually linking interdependent primary and support activities to deliver better service at destinations that ultimately positively impact the socio-culture and economy (Cheng et al., 2020).

It is also evident that shifting transformation and convergence of technologies within the digital media atmosphere have started to counteract the harms of social media exposure (Gretzel, 2019). The mutating relations of social community, power and practice through digital systems interpenetrate all functional aspects of event tourism in general and the upturn of the sharing economy in particular. The sharing economy, established by the peer-to-peer platforms for

Table 1.1 Recognized established models in tourism

Model and Theory	Description	Sources
The Theory of Planned Behavior (TPB)	The Theory of Planned Behavior is considered as one of the recognized frameworks initially developed by Ajzen (1985, 1991) applied in various research studies to examine the behaviour and intentions of tourists from different perspectives. This constructive theory claims that behavioural intention is driven by attitude toward behaviour, subjective norms and perceived behavioural control (Ajzen, 1991). According to the TPB emerged as a viable theoretical framework for evaluating the motivations of individuals who attended the tourism events and festivals.	Tölkes and Butzmann, 2018; Chung et al., 2018; Ting et al., 2017; Alonso et al., 2015; Fragkogianni, 2018; Kaplan et al., 2015; Quintal et al., 2010
The Technology Acceptance Model (TAM)	TAM is a well-developed model that has been applied to examine why many tourists are not interested in using smartphones/technology from the service sector firms. Many researchers also adopted the technology acceptance model (TAM) to understand the intentions of tourists toward technology use in the virtually driven tourism environment. In some specific tourism researches, TAM is integrated with other theories to modify the theoretical framework for examining user's intention, visitors' experience and behavioural intentions (Huang et al., 2013), online purchase travel intention, emotional drivers toward the use of digital channels, attitudes towards using franchise intranet.	Kwon et al., 2013; Im and Hancer, 2014; Huang et al., 2015; Amaro and Duarte, 2015; Straker and Wrigley, 2016; Park et al., 2014
Unified Theory of Acceptance and Use of Technology (UTAUT)	The Unified Theory of Acceptance and Use of Technology (UTAUT) model is one of the most frequent and recent consumer behaviour models to be utilized for analysing the tourism industry. This model developed by Venkatesh, Morris, Davis and Davis (2003) have already been applied in tourism technology research and received wide acceptability in academia. It has been established the fact that effort expectancy, performance expectancy, social influence and facilitating constituents are the crucial factors ascertaining user adoption (Venkatesh et al., 2003). The UTAUT2 model, which is an improved version of the UTAUT model, has also been widely adopted in the tourism industry.	Bakar et al., (2020); Moghavvemi and Salleh, 2014; Fonseca et al., (2020)
Social Cognitive Theory (SCT)	In recent years, social cognitive theory (SCT), which originates from social learning theory, has gained widespread recognition in technology-focused tourism research as a comprehensive conceptual framework for assessing human behaviour (Bandura, 1986; Lu et al., 2015). SCT has been widely utilized in the study to evaluate technology adoption as well as sustainable tourism (Afolabi et al., 2020; Font et al., 2016). In mobile travel apps usage, expectations towards outcome are considered as a significant motivating predictor of behaviour, which confirms the propositions obtained from social cognitive theory (SCT) (Lu et al., 2015).	Afolabi et al., 2020; Font et al., 2016; Jepson and Ryan, 2018
Innovation Diffusion Theory (IDT)	Recently, IDT has received widespread recognition in examining consumers' intention towards tourism (Agag and El-Masry, 2016; Gu et al., 2019). IDT is being utilized to understand the travellers' behaviour better morally exchanging across the population, highlighting the association between relatively stagnant tourism innovation and the dissemination of innovation in the tourism industry (Bilgihan and Wang, 2016). IDT presents an intriguing model that helps to understand tourists' intentions towards joining the online travel community, which notably affects purchasing intention and generates positive word of mouth (Agag and El-Masry, 2016).	Gu et al., 2019; Agag and El-Masry, 2016; Bilgihan and Wang, 2016

collaborative, customized consumption, enhances service marketing practices at a standard level (Guttentag, 2015). In particular, the sharing economy is connected with the growing importance of co-creation in the tourism business in the form of system, experience sharing and value delivery (Campos et al., 2018). One of the situational facts is that immersive technologies like virtual reality or augmented reality have been applied in the tourism sector to enhance the on-site visitor experience. Such kinds of technologies create various layers of media content to imbrue tourists in destinations and attractions (Bec et al., 2021; Scarles et al., 2020). Promoting through virtual reality creates potential tourists to interact with attractive spots online well in advance before making a decision (Yung and Khoo-Lattimore, 2019). Furthermore, telepresence tourism propagates alternative opportunities for greater engagement in entertainment programs for participants who are mobility restricted (Scarles et al., 2020).

Digital Innovation in Tourism Events for Competitive Advantages

In the era of the sophisticated business world, it has been evident that modern tourism enterprises may not sustain unless they are capable of ensuring digital innovation. The demand for further continuous growth, the need to cope with increasing competition, complexity and the frenetic motion of change eventually forces enterprises to meet the innovative challenge (Aldebert et al., 2011). The evolution of e-commerce has gained acceptability as a strategic constituent of achieving competitive advantage in all branches of the tourism industry (Kim et al., 2009). Okumus (2013) argued that tourism enterprises could enhance knowledge management and gain distinctive competitive strength through various tools of IT. Tourism enterprises need to create a favourable organizational structure and culture along with motivated, trained employees to manage knowledge in a systematic manner through IT mechanisms in order to enhance value co-creation and strategic benefits (Buonincontri and Micera, 2016; Cabiddu et al., 2013).

Digital innovation is not just about entering new diverse markets, but it can also enhance the alternative way tourism services are provided to the matured ones. Matured tourism enterprises should have the dynamic capability to ensure digital innovation for the competitive reason (Arvidsson et al., 2014). Practically, tourism events are considered as a strategic approach (Todd et al., 2017), and competitiveness is driven by perfect resources acquisition, resources integration and resources deployment at the right time and place (McKercher, 2016; Tanford and Jung, 2017). In order to enhance the competitive advantage, managers of tourism enterprises can adapt their knowledge, technological skills and market experience to bring value, generating new dimensions to tourism events (Kelly and Fairley, 2018). Event tourism enterprises are actively seeking to manage networked operating mechanisms to enhance the experience of attendees (Winkle et al., 2021). Forces and actors in the business environment are changing their role and characteristics; that is the way it is undoubtedly important for tourism enterprises to focus on innovative service development in social media by incorp-orating sophisticated technologies (Sigala, 2019). From a tourism industry perspective, it can be said that there is a positive correlation between market performance and innovative technology adoption (Gomezelj, 2016). The com-petitiveness of any event tourism enterprise of tourist location mostly depends on its capability to adopt mobile technologies to create a favourable atmosphere that mobilizes such a sensitive service-driven tourism sector (Keen and Mackintosh, 2001). In the matured and established products, competitive market growth is not only derived from low-price tourism service offerings, but also to some con-siderable extent non price factors such as quality, design, customization, just-in-time delivery and customer service. These non price factors can be managed by incorporating digital innovation in tourism sectors in a synergetic fashion.

From a strategic standpoint, digital innovation is as important as service innovation in a competitive business atmosphere. If a tourism enterprise has knowledge, resources and know-how to create, communicate and deliver some-thing better value than surrounding competitors, or being able to ensure that no other competitors can imitate such things, could be a great source of com-petitive advantage. Small tourism firms face problems coping with the competi-tive situation due to limited financial resources and technological know-how. Alternatively, small firms can go for strategic alliances with other firms to enhance their digital innovation capability for survival in the competitive battle-field. Not all industries offer the same market opportunities and economic returns. The fundamental competitive factors that drive a tourism enterprise's ability to generate above-average economic performance are: (1) bargaining power of customers; (2) bargaining power of suppliers; (3) rivalry among existing competitors; (4) threat from new competitors; and (5) threat of sub-stitute products (Cimbaljević et al., 2021). The conventional analysis suggests selecting industries within these criteria; however, it is worth mentioning that enterprises must rely on digital innovation, positioning strategies and how they manage organizational employees and culture to gain a competitive advantage. Traditional success factors (technology, product or services, investment, econ-omies of scale) still do assure competitiveness, but to a lesser degree now than in the past. The culture and dynamic capabilities of the organization derive from how people are managed towards achieving goals. However, any source of competitiveness erodes away by the changes of time. As the business environ-ment changes over time, the need for continuous digital innovation and prompt response to market dynamics and technological changes virtually requires a well-trained workforce that delivers outstanding performance.

Firms will have to find scope to gain and sustain competitive advantage to maintain event tourism industry attractiveness and superior performance. Competition has intensified across all industries and very few tourism business environments can guarantee secure economic returns. Many tourism enterprises have achieved remarkable growth in recent decades and expanded quickly through further growth and diversification. However, many others have lost their competitive position through the effortless growth they have enjoyed. Nowadays, tourism enterprises around the globe are faced with incremental growth rate and increased diverse competition both locally and globally. The competitive advantage thus becomes a much more vital issue to ensure the business performance and survival of these enterprises, particularly with the introduction of digital innovation, which may erode the competitive advantage of many firms. Competitive advantage is the ability of the tourism firm to outperform better than rivals on the primary performance goal –profitability. It is true that economic return should only be viewed as a long-term achievement of competitive advantage as an investment in research, technology adoption, customer satisfaction etc. to build market share may affect short term rents but will offer the firm with long term competitive advantage (Ismail et al., 2013). The remarkable growth of digital technology and the emphasis on digital innovation worldwide can certainly affect the event tourism industry structure, market life cycle and competitive advantage. Digital innovation drives business growth as it has opened up huge innovative opportunities for tourism firms that move quickly to establish digital channels and create significant threats to others who fail to realize the strategic importance of technological resources. From corporate-level support, actors in the industry can apply ICT to build and maintain a competitive edge, obtain realistic information capabilities and actively concentrate on their key success factors to amplify the value of the supply chain through the collaboration of interdependent organizations.

Policymakers of the event tourism industry may adopt Internet-based decision support systems at all levels that are useful to enhance the favourable industrial atmosphere and competitive economic performance (Bilgihan and Wang, 2016). There is a number of macro-level forces (political, legal, economic, technological and socio-cultural) that are very much associated with the tourism environment. Apart from that, micro-level diverse competitive five forces also drive tourism functions. There is a certain degree of uncertainties, risks and new challenges associated with the ever-changing industry environment. To mitigate risks and face upcoming challenges, policymakers must systematically apply all the crucial management functions and follow a four-stage approach to gain a competitive advantage (Figure 1.1). First, apply appropriate models to evaluate the situational reality of techno-driven event tourism industry; second, formulate and implement competitive strategies by the configuration of techno and economic resources in a synergetic fashion; third, enhance distinctive competence for delivering value to tourists better than competitors; and finally evaluate and ensure industry performance from stakeholders' expectations point of view.

Conclusion

This chapter has evaluated the dynamics of event tourism in relation to digital innovation with the advent of technological advancements. IT impacts the event tourism industry structure, and it can create and provide event management firms with a huge competitive advantage and above-average performance. IT brings extensive benefits to event tourism firms and provides them with a competitive advantage if firms adopt digital innovation. Event tourism firms that do not

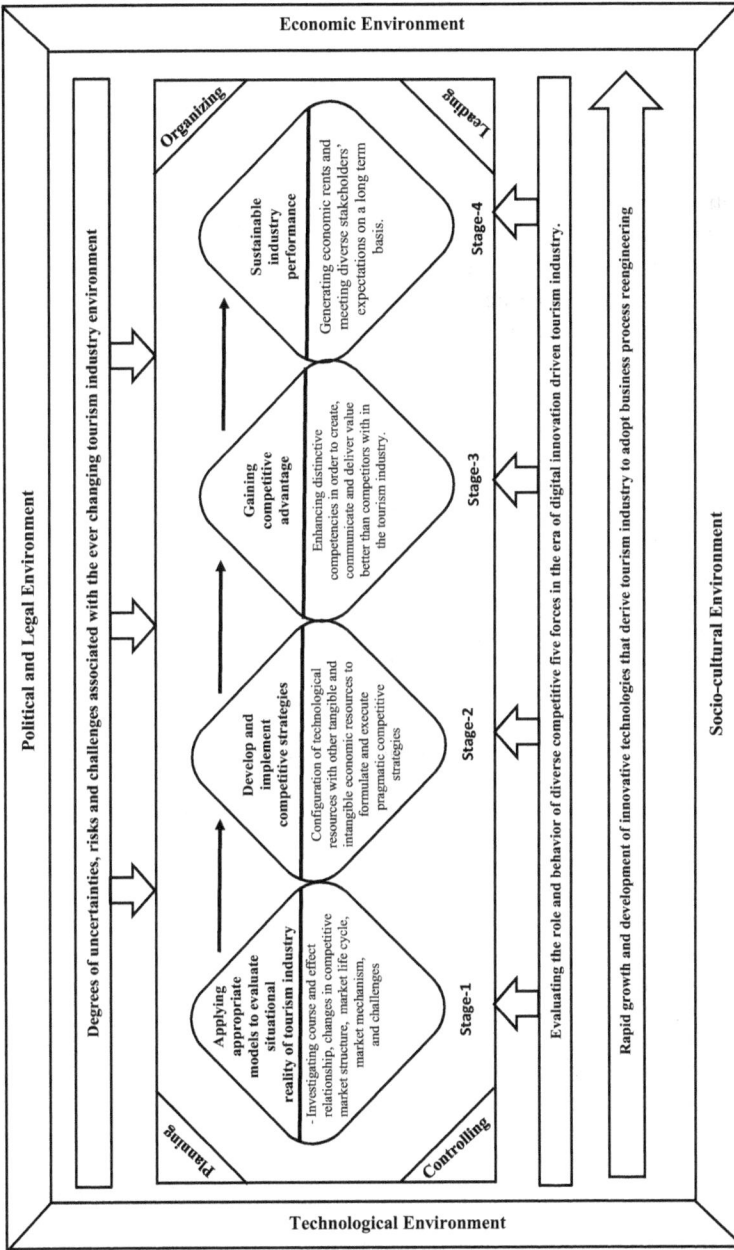

Figure 1.1 A four-stage approach to gain a competitive advantage

Source: Developed by the authors

embrace sophisticated technology will be faced with threats from other firms that have successfully adopted the digital innovation. The emerging competitive market also means that event tourism firms have no choice but to adopt the technology or risk losing out from the market. Although there are many challenges and risks involved in the adoption of digital innovation, event tourism firms should take precautions to deal with them head-on and should not be afraid of them. The success of many Web-based companies has demonstrated that the benefits of IT can outweigh the costs and challenges. The future of digital innovation through adoption of smart technology is seen to be bright due to changing tourists' behaviour towards a comfortable smart purchase, increasing awareness and use of the Internet. The benefits and opportunities of technology innovation (cost-saving and distinctive value creation) and the threats for not adopting them are far too great to ignore the new channel. Hence event tourism firms should pack their bags and set off to the future of smart IT.

References

Afolabi, O.O., Ozturen, A., and Ilkan, M. (2020). Effects of privacy concern, risk, and information control in a smart tourism destination. *Economic Research-Ekonomska Istrazivanja*. https://doi.org/10.1080/1331677X.2020.1867215.

Agag, G., and El-Masry, A.A. (2016). Understanding consumer intention to participate in online travel community and effects on consumer intention to purchase travel online and WOM: An integration of innovation diffusion theory and TAM with trust. *Computers in Human Behavior*, 60, 97–111.

Ajzen, I. (1985) From Intentions to Actions: A Theory of Planned Behavior. In J. Kuhl and J. Beckmann (Eds.), *Action Control. SSSP Springer Series in Social Psychology*. Berlin, Heidelberg: Springer, pp. 11–39.

Ajzen, I. (1991). The theory of planned behavior. *Organizational Behavior and Human Decision Processes*, 50(2), 179–211.

Aldebert, B., Dang, R.J., and Longhi, C. (2011). Innovation in the tourism industry: The case of Tourism@. *Tourism Management*, 32(5), 1204–1213.

Alonso, A.D., Sakellarios, N., and Cseh, L. (2015). The theory of planned behavior in the context of a food and drink event: A case study. *Journal of Convention and Event Tourism*, 16(3), 200–227.

Amaro, S., and Duarte, P. (2015). An integrative model of consumers' intentions to purchase travel online. *Tourism Management*, 46, 64–79.

Arvidsson, V., Holmström, J., and Lyytinen, K. (2014). Information systems use as strategy practice: A multi-dimensional view of strategic information system implementation and use. *Journal of Strategic Information Systems*, 23(1), 45–61.

Bakar, N.A., Aliff, N., Yusoff, A.M., and Rahim, M.A. (2020). Travel mobile applications: The use of Unified Acceptance Technology Model. *International Journal of Innovative Technology and Exploring Engineering*, 9(3), 3118–3121.

Bandura, A. (1986). *Social Foundations of Thought and Action: Social Cognitive Theory*. Englewood Cliffs, NJ: Prentice Hall.

Bec, A., Moyle, B., Schaffer, V., and Timms, K. (2021). Virtual reality and mixed reality for second chance tourism. *Tourism Management*, 83. https://doi.org/10.1016/j.tourman.2020.104256

Berne, C., Garcia-Gonzalez Margarita, M., and Mugica, J. (2012). How ICT shifts the power balance of tourism distribution channels. *Tourism Management*, 33(1), 205–214.

Bilgihan, A., and Wang, Y. (2016). Technology induced competitive advantage: A case of US lodging industry. *Journal of Hospitality and Tourism Technology*, 7(1), 37–59.

Breukel, A., and Go, F.M. (2009). Knowledge-based network participation in destination and event marketing: A hospitality scenario analysis perspective. *Tourism Management*, 30(2), 184–193.

Buonincontri, P., and Micera, R. (2016). The experience co-creation in smart tourism destinations: A multiple case analysis of European destinations. *Information Technology and Tourism*, 16(3), 285–315.

Cabiddu, F., Lui, T.W., and Piccoli, G. (2013). Managing Value Co-Creation In The Tourism Industry. *Annals of Tourism Research*, 42, 86–107.

Campos, A.C., Mendes, J., do Valle, P.O., and Scott, N. (2018). Co-creation of tourist experiences: A literature review. *Current Issues in Tourism*, 21(4), 369–400.

Cheng, M., Houge Mackenzie, S., and Degarege, G.A. (2020). Airbnb impacts on host communities in a tourism destination: an exploratory study of stakeholder perspectives in Queenstown, New Zealand. *Journal of Sustainable Tourism*. https://doi.org/10.1080/09669582.2020.1802469

Chung, J.Y., Kim, J.S., Lee, C.K., and Kim, M.J. (2018). Slow-food-seeking behaviour, authentic experience, and perceived slow value of a slow-life festival. *Current Issues in Tourism*, 21(2), 123–127.

Cimbaljević, M., Stankov, U., Demirović, D., and Pavluković, V. (2021). Nice and smart: creating a smarter festival–the study of EXIT (Novi Sad, Serbia). *Asia Pacific Journal of Tourism Research*, 26(4), 415–427.

del Vecchio, P., Secundo, G., and Passiante, G. (2018). Modularity approach to improve the competitiveness of tourism businesses: Empirical evidence from case studies. *EuroMed Journal of Business*, 13(1), 44–59.

Fonseca, D., Reis, J.L., Teixeira, S., and Peter, M.K. (2020). Mobile applications at music festivals in Portugal. *Smart Innovation, Systems and Technologies*, 167, 190–201.

Font, X., Garay, L., and Jones, S. (2016). A Social Cognitive Theory of sustainability empathy. *Annals of Tourism Research*, 58, 65–80.

Fragkogianni, M. (2018). Formulating event loyalty in tourism: Lessons from the world travel market. *Event Management*, 22(1), 37–47.

Garbelli, M., Adukaite, A., Cantoni, L., and Adukaite, A. (2017). Value perception of world heritage sites and tourism sustainability matters through content analysis of online communications: The case of Victoria Falls world heritage site. *Journal of Hospitality and Tourism Technology*, 8(3), 417–431.

George, B., and Paul, J. (2019) (Eds.). *Digital Transformation in Business and Society: Theory and Cases*. Cham: Palgrave Macmillan.

Goldkuhl, G. (2012). Pragmatism vs interpretivism in qualitative information systems research. *European Journal of Information Systems*, 21(2), 135–146.

Gomezelj, D.G. (2016). A systematic review of research on innovation in hospitality and tourism. *International Journal of Contemporary Hospitality Management*, 28(3), 516–558.

Gretzel, U. (2019). The role of social media in creating and addressing overtourism. In R. Dodds and R. Butler (Eds.), *Overtourism: Issues, Realities and Solutions*. Berlin, Boston: De Gruyter Oldenbourg, 62–75.

Grissemann, U.S., and Stokburger-Sauer, N.E. (2012). Customer co-creation of travel services: The role of company support and customer satisfaction with the co-creation performance. *Tourism Management*, 33(6), 1483–1492.

Grossman, R. (2016). The Industries That Are Being Disrupted the Most by Digital. *Harvard Business Review*. Retrieved from: https://bit.ly/3neIfgY (accessed 19 October 2021).

Gu, D., Khan, S., Khan, I.U., and Khan, S.U. (2019). Understanding mobile tourism shopping in Pakistan: An integrating framework of innovation diffusion theory and technology acceptance model. *Mobile Information Systems*, https://doi.org/10.1155/2019/1490617.

Gursoy, D., Nunkoo, R., and Yolal, M. (2020) (Eds.). *Festival and Event Tourism Impacts*. Abingdon: Routledge.

Guttentag, D. (2015). Airbnb: Disruptive innovation and the rise of an informal tourism accommodation sector. *Current Issues in Tourism*, 18(12), 1192–1217.

Hannam, K., Butler, G., and Paris, C.M. (2014). Developments and key issues in tourism mobilities. *Annals of Tourism Research*, 44(1), 171–185.

Higgins-Desbiolles, F. (2018). Event tourism and event imposition: A critical case study from Kangaroo Island, South Australia. *Tourism Management*, 64, 73–86.

Hinings, B., Gegenhuber, T., and Greenwood, R. (2018). Digital innovation and transformation: An institutional perspective. *Information and Organization*, 28(1), 52–61.

Huang, J.M., Ho, T.K., Liu, Y.C., and Lin, Y.H. (2015). A discussion on the user intention of golfers toward golf GPS navigation. *Journal of Hospitality and Tourism Technology*, 6(1), 26–39.

Huang, Y.C., Backman, S.J., Backman, K.F., and Moore, D.W. (2013). Exploring user acceptance of 3D virtual worlds in travel and tourism marketing. *Tourism Management*, 36, 490–501.

Im, J.Y., and Hancer, M. (2014). Shaping travelers' attitude toward travel mobile applications. *Journal of Hospitality and Tourism Technology*, 5(2), 177–193.

Ismail, A.F., Zorn, S.F., Chern Boo, H., Murali, S., and Murphy, J. (2013). Information technology diffusion in Malaysia's food service industry. *Journal of Hospitality and Tourism Technology*, 4(3), 200–210.

Jepson, A., and Ryan, W.G. (2018). Applying the motivation, opportunity, ability (MOA) model, and self-efficacy (S-E) to better understand student engagement on undergraduate event management programs. *Event Management*, 22(2), 271–285.

Jin, S. V., and Phua, J. (2016). Making reservations online: The impact of consumer-written and system-aggregated User-Generated Content (UGC) in travel booking websites on consumers' behavioral intentions. *Journal of Travel and Tourism Marketing*, 33(1), 101–117.

Kaplan, S., Manca, F., Nielsen, T.A.S., and Prato, C.G. (2015). Intentions to use bike-sharing for holiday cycling: An application of the Theory of Planned Behavior. *Tourism Management*, 47, 34–46.

Keen, P.G., and Mackintosh, R. (2001). *The Freedom Economy: Gaining the mcommerce Edge in the Era of the Wireless Internet*. London: McGraw-Hill Professional.

Kelly, D.M., and Fairley, S. (2018). What about the event? How do tourism leveraging strategies affect small-scale events? *Tourism Management*, 64, 335–345.

Kim, Y.G., Eves, A., and Scarles, C. (2009). Building a model of local food consumption on trips and holidays: A grounded theory approach. *International Journal of Hospitality Management*, 28(3), 423–431.

Kwon, J.M., Bae, J., and Blum, S.C. (2013). Mobile applications in the hospitality industry. *Journal of Hospitality and Tourism Technology*, 4(1), 81–92.

Lamsfus, C., Wang, D., Alzua-Sorzabal, A., and Xiang, Z. (2015). Going mobile: Defining context for on-the-go travelers. *Journal of Travel Research*, 54(6), 691–701.

Lee, S.S., Boshnakova, D., and Goldblatt, J. (2016). *The 21st Century Meeting and Event Technologies: Powerful Tools for Better Planning, Marketing, and Evaluation.* Abingdon: Apple Academic Press.

Leung, D., Law, R., van Hoof, H., and Buhalis, D. (2013). Social Media in Tourism and Hospitality: A Literature Review. In *Journal of Travel and Tourism Marketing*, 30(1–2), 3–22.

Li, F. (2020). The digital transformation of business models in the creative industries: A holistic framework and emerging trends. *Technovation*, 92(93), 102012.

Li, J., Pearce, P.L., and Low, D. (2018). Media representation of digital-free tourism: A critical discourse analysis. *Tourism Management*, 69, 317–329.

Lu, J., Mao, Z., Wang, M., and Hu, L. (2015). Goodbye maps, hello apps? Exploring the influential determinants of travel app adoption. *Current Issues in Tourism*, 18(11), 1059–1079.

McCabe, S., Sharples, M., and Foster, C. (2012). Stakeholder engagement in the design of scenarios of technology-enhanced tourism services. *Tourism Management Perspectives*, 4, 36–44.

McKercher, B. (2016). Towards a taxonomy of tourism products. *Tourism Management*, 54, 196–208.

Mitchell, V.W., Schlegelmilch, B.B. and Mone, S.D. (2016). Why should I attend? The value of business networking events. *Industrial Marketing Management*, 52, 100–108.

Moghavvemi, S., and Salleh, N.A.M. (2014). Effect of precipitating events on information system adoption and use behaviour. *Journal of Enterprise Information Management*, 27(5), 599–622.

Murphy, H.C., Chen, M.M., and Cossutta, M. (2016). An investigation of multiple devices and information sources used in the hotel booking process. *Tourism Management*, 52, 44–51.

Nambisan, S., Lyytinen, K., Majchrzak, A., and Song, M. (2017). Digital innovation management: Reinventing innovation management research in a digital world. *MIS Quarterly: Management Information Systems*, 41(1), 223–238.

Neuhofer, B., Buhalis, D., and Ladkin, A. (2013). Co-creation Through Technology: Dimensions of Social Connectedness. In Z. Xiang and I. Tussyadiah (Eds.), *Information and Communication Technologies in Tourism 2014* (pp. 339–352). Cham: Springer.

Okumus, F. (2013). Facilitating knowledge management through information technology in hospitality organizations. *Journal of Hospitality and Tourism Technology*, 4(1), 64–80.

Park, K., Lee, S.S., and Khan, M.A. (2014). Exploring the impact of franchise support on franchisee acceptance of intranet in quick service restaurant (QSR) franchise system. *Journal of Hospitality and Tourism Technology*, 5(2), 143–159.

Quintal, V.A., Lee, J.A., and Soutar, G.N. (2010). Risk, uncertainty and the theory of planned behavior: A tourism example. *Tourism Management*, 31(6), 797–805.

Raj, R., and Musgrave, J. (2009) (Eds.). *Event Management and Sustainability.* Wallingford: CABI Publishing.

Scarles, C., Even, S. van, Klepacz, N., Guillemau, J.Y., and Humbracht, M. (2020). Bringing the outdoors indoors: Immersive experiences of recreation in nature and coastal environments in residential care homes. *E-Review of Tourism Research*, 17(5), 706–721.

Sigala, M. (2019). The bright and the dark sides of social media in tourism experiences, tourists' behavior, and well-being. In D.J. Timothy (Ed.), *Handbook of Globalisation and Tourism*. Cheltenham: Edward Elgar, pp. 247–259.

Sigala, M., and Chalkiti, K. (2014). Investigating the exploitation of web 2.0 for knowledge management in the Greek tourism industry: An utilisation-importance analysis. *Computers in Human Behavior*, 30, 800–812.

Sigala, M., Christou, E., and Gretzel, U. (2016). Social media in travel, tourism and hospitality: Theory, practice and cases. In E. Christou and M. Sigala (Eds.), *Social Media in Travel, Tourism and Hospitality: Theory, Practice and Cases*. Abingdon: Routledge.

Straker, K., and Wrigley, C. (2016). Translating emotional insights into digital channel designs: Opportunities to enhance the airport experience. *Journal of Hospitality and Tourism Technology*, 7(2), 135–157.

Tanford, S., and Jung, S. (2017). Festival attributes and perceptions: A meta-analysis of relationships with satisfaction and loyalty. *Tourism Management*, 61, 209–220.

Ting, H., Mering, M.W., Adruce, S.A.Z., and Memon, M.A. (2017). Intention to attend the Rainforest World Music Festival: Local visitor perspectives. *Tourism, Culture and Communication*, 17(2), 119–129.

Todd, L., Leask, A., and Ensor, J. (2017). Understanding primary stakeholders' multiple roles in hallmark event tourism management. *Tourism Management*, 59, 494–509.

Tölkes, C., and Butzmann, E. (2018). Motivating pro-sustainable behavior: The potential of green events-A case-study from the Munich Streetlife Festival. *Sustainability*, 10(10), 3731.

van Nuenen, T. (2016). Here I am: Authenticity and self-branding on travel blogs. *Tourist Studies*, 16(2), 192–212.

van Nuenen, T. (2019). Algorithmic authenticity: Sociotechnical authentication processes on online travel platforms. *Tourist Studies*, 19(3), 378–403.

Venkatesh, V., Morris, M. G., Davis, G. B., and Davis, F. D. (2003). User acceptance of information technology: Toward a unified view. *MIS Quarterly: Management Information Systems*, 27(3), 425–478.

Willis, C., Ladkin, A., Jain, J., and Clayton, W. (2017). Present whilst absent: Home and the business tourist gaze. *Annals of Tourism Research*, 63, 48–59.

Winkle, C.M. Van, MacKay, K.J., and Halpenny, E. (2021). Information and communication technology and the festival experience. In J. Mair (Ed.), *The Routledge Handbook of Festivals*. Abingdon: Routledge, 254–262.

Wozniak, T., Stangl, B., Schegg, R., and Liebrich, A. (2017). The return on tourism organizations' social media investments: Preliminary evidence from Belgium, France, and Switzerland. *Information Technology and Tourism*, 17(1), 75–100.

Xu, F., Huang, S., and Li, S. (2019). Time, money, or convenience: What determines Chinese consumers' continuance usage intention and behavior of using tourism mobile apps? *International Journal of Culture, Tourism, and Hospitality Research*, 13(3), 288–302.

Yung, R., and Khoo-Lattimore, C. (2019). New realities: A systematic literature review on virtual reality and augmented reality in tourism research. *Current Issues in Tourism*, 22(17), 2056–2081.

Part II

Cases from Asia

2 Events at Nature-based Destinations of Bangladesh

Use of Information and Communication Technology in Marketing

Md. Wasiul Islam

Introduction

The digital information age has significant impacts on global tourism (Aramendia-Muneta and Ollo-López, 2013; COMCEC, 2015; Gössling, 2020, 2021; Khan and Hossain, 2018). Nowadays, the application of Information and Communication Technology (ICT) has profound implications on the growth and development of tourism and tourism destinations. ICT is intimately functioning as a driver of sustainable tourism development by fostering the opportunities of economic, socio-cultural and environmental benefits especially for developing economies (Abdullayeva, 2019; Gössling, 2021; Januszewska et al., 2015; Shanker, 2008; Vasiliki and Maria, 2013). Advanced ICT and other technologies are also used in the hospitality and tourism sector to create new tourism products and services as well as to move their businesses into the globalized market (Abdullayeva, 2019), which facilitates the nexus between developed and developing economies by increasing the access of tourism products and services.

Different festivals, special events and exhibitions are the cultural resources of a tourism destination including nature-based destinations which are able to attract visitors and creates a new economy (Backman et al., 1995; Stratigea et al., 2008). The event industry is a fast-growing and vibrant sector which strongly influences the tourism sector (van Eck, 2018; Patwary et al., 2020). These events are known as a focal part of tourism development in many destinations around the world (Dwyer and Wickens, 2011) and form a distinct type of tourism (i.e., event tourism) (Getz, 2008). Event tourism is a sub-section of tourism which is evolved from tourism studies and event studies (Getz and Page, 2016).

Nowadays, the application of ICT is very common in the tourism sector, including event tourism (COMCEC, 2015; Farkhondehzadeh et al., 2013; Tahayori and Moharrer, 2006). The application and integration of ICT in the tourism sector is considered as part and parcel for the success of this sector including all types of tourism enterprises. The main function of ICT is to establish communication between supply and demand. From demand side, potential

DOI: 10.4324/9781003271147-5

tourists can access ICT facilities to be familiar about the attractions and services of a tourist destination from any corner of the world at any time, mainly through the World Wide Web and social media. From supply side, it serves the information from the service providers (e.g., accommodation, cuisines, guiding, and so on) to the potential tourists through developing new and innovative marketing channels (Stratigea and Giaoutzi, 2017; Tahayori and Moharrer, 2006).

The study has deliberately focused on Bangladesh which is culturally, ecologically, historically and politically very rich. There are several festivals, special events and exhibitions taking place at different destinations including nature-based destinations. These events are organized to create a positive image of these destinations and to facilitate socio-economic development of the destination communities as well as other business agents. The extant literature shows there is no specific research on such events at nature-based destinations and the contributions of ICT in marketing these events. Therefore, the chapter contains the study aiming to investigate the event tourism at the nature-based tourism destinations of Bangladesh and the contributions of ICT in their marketing, mainly promotion of these events. The study is mainly based on systematic literature review. Extant literature has been collected from different journal articles, books, newspapers, online database, blogs, videos and personal communication with some resource persons.

Events at Nature-based Destinations of Bangladesh

Bangladesh is popular for various colourful cultural, religious and national festivals taking place all year round. The country is the abode of several ethnic tribes along with the mainstream population who are very rich in their cultural and religious festivals. Therefore, it is known as a festive country. These festive events take place in different settings of Bangladesh including rural and urban areas, nature-based and built environments. After an extensive extant literature review it may be concluded that there are not many major events taking place particularly at the nature-based destinations. Most of these events are cultural events, with more specifically religious events celebrated annually mainly in the southern part of Bangladesh. The literature review suggested the following notable events occur at the nature-based destinations in the country (Table 2.1).

Rashmela

Rashmela is a religious festival which is the main and biggest event in the nature-based tourism destinations of Bangladesh. It is an annual festival that takes place at Dubla island (more specifically at Alorkol) of the Sundarbans mangrove forest which is the world's single largest mangrove forest situated in the southwestern part of Bangladesh (72%) and West Bengal of India (28%). This unique mangrove forest plays a very crucial role in terms of socio-cultural, economic and environmental factors. Rashmela is one of the instances of not only socio-cultural factors but also economic as well as environmental factors.

Table 2.1 Events at nature-based destinations of Bangladesh: at a glance

Events	Destinations	Main features
Rashmela	Dubla island of the Sundarbans Madhabpur, Moulvibazar Adampur and Jayashree, Habiganj Kuakata see beach, Patuakhali Kantojir temple, Dinajpur	• Hindu religion based annual festival • Commemorate the spiritual *Rasalila* (love-play) between Radha and Krishna at Vrindavan • Held full moon of *Kartik/ Agrahayan* of Bengali calendar
Honey hunting	All over the Sundarbans	• Popular for adventurous tourists • A cultural and livelihood option • Starts on 1 April and ends on 30 June each year
Baishabi festival	Chottogram Hill Tracts	• Religion based annual festival • Celebrated by Chakma, Tanchangya, Tripura and Marma ethnic tribes • Celebrated during the Bengali New Year (14 April of each year)
Boat racing	Several rivers of all over the country	• A riverine cultural annual festival that generally occurs during the autumn • Annual boat race festival on the Buriganga river at Dhaka
Full moon festival	Shuvo Sondha beach, Barguna	• Newly formed beach • Lot of birds along with other wildlife are available • Sun set is clearly visible here
Wangala festival	Greater Mymensingh	• Thanksgiving and major religious festival of the Garo ethnic tribe • Offer their newly harvested crops to their Gods • Occurs annually in the month of *Kartik*
Shiv Chaturdashi fair	Shiva temple, Chandranath hill, Chottogram	• Hindu religious annual festival • One of the oldest and major cultural and religious fairs in the sub-continent • Starts on 14th night of *Falgun*
Events and festivals at beaches	Several points of Cox's Bazar, Kuakata, St. Martin, and Patenga sea beach	• Various national and cultural festivals take place all over the year • Other events like meetings, conferences, etc. take place
Holiday-based events	All over the country	• Holidays like two Eid festivals, Pujas, and other vacations/ holidays

The festival is organized in the full moon of November (the month of *Kartik/ Agrahayan* in the Bengali calendar depending on lunar month) every year for three days. Rashmela is a Hindu religion-based festival to commemorate the divine love of Hindu God Lord Sri Krishna and Radha. More specifically, the people of Bishnupriya community of Munipuri tribal community celebrate this festival as their main festival. The name of the festival was evolved from the spiritual *Rasalila* (love-play) between Radha and Krishna at Vrindavan. However, besides these Hindu pilgrims, people of different religions (around 40,000–50,000 tourists) from different countries also visit this island along with other destinations of the Sundarbans during this time.

Honey hunting

The Sundarbans is bestowed with a variety of plants. Some of these plants have nectars in their flowers which attract bees to collect it. Among them *Khalsi* (*Aegiceras corniculatum*), *Goran* (*Ceriops decandra*) and *Passur* (*Xylocarpus mekongensis*) flowers are most popular with the bees. There are enormous honey hives in the Sundarbans. On average, over 200 tons of honey is collected from the Sundarbans along with a smaller amount (50 tons) of beeswax. There is a group of people who are called *Mouwali* (honey collectors) in Bengali. Their honey collection art and activities are also known as one of the tourism attractions of the Sundarbans particularly for the adventurous tourists. Some of the tour operators arrange honey-hunting package tours during this season. Some tourists also visit the Sundarbans by their own arrangements. Honey hunting officially starts on 1 April and ends on 30 June each year. The villagers living around the Sundarbans used to collect wild honey during this period with the permission of the Forest Department. Tourists having special protection from bees can follow the honey collectors to gather unique experiences of collecting honey from the wild environment which is full of various challenges and dangers.

Baishabi festival

This festival is celebrated by the several ethnic tribes of Chottogram Hill Tracts (CHT) situated in the southeastern part of Bangladesh. The colourful festival is celebrated focusing on the Bengali New Year (14 April of each year) to say farewell to the past year and welcome the New Year. The festival is known by various names according to the ethnic tribal sects of the region. The Chakma and Tanchangya people call it *Biju* (*Bizu* or *Bishu*), the Tripura *Boisu* and Marma *Sangrai*. These three festivals are collectively known as *Baishabi*. Though there are some ritual differences among these three festivals, the main spirit is the same. Moreover, different ethnic events such as Chakma dance, Marma Chata dance, *Pani khela* (water festival) along with other sport events, Tripura *Goraiya* dance, fairs, and so on are arranged during the festival highlighting their distinctive cultural features. The water festival is one of the most popular events of the Marma community during the season. All these three festivals (*Baishabi*) attract tourists from all over the country and even foreign tourists visit during this season. Three districts of CHT – namely Rangamati, Bandarbans and Khagrachari – are usually popular tourist destinations of Bangladesh. The demand to visit these destinations has increased due to these colourful festivals. Moreover, nowadays these festivals are growing more popular due to digital marketing through social media and other print and electronic media. Many tour operators offer special tour packages on this auspicious occasion.

Boat racing

Bangladesh is the world's largest delta having more than 230 rivers. The culture and livelihoods of Bangladeshi people are greatly influenced by these rivers. There are several popular boat race (*Nouka Baich* in Bengali) festivals in riverine Bangladesh which are treated as heritage, social events and elements of folk culture which are great sources of entertainment for both the rowers and visitors. Millions of visitors gather on the riversides to enjoy these boat races. Generally, these boat races take place at the rural areas during the autumn season (*Bhadra* and *Ashwin* months of Bengali calendar). Different types of boats are used in such races depending on the places. The historical annual boat race festival on the Buriganga river at Dhaka is the most popular one. Similar types of boat racing are organized all around the country such as those at Sirajganj, Khulna, Gopalganj, Shariatpur, Norail, Pabna, Magura, Brahmanbaria, Madaripur, Jessore, Sylhet, Tangail, Munshiganj, Narayanganj and Narsingdi districts, among others. However, this fascinating heritage is currently facing threats due to various causes including funding, organizing, managing, river pollution, mechanized and modern lifestyles, and so on.

Full moon festival

Full moon festival is an annual event taking place at Shuvo Sondha beach located at Taltoli sub-district of Barguna district. The three main rivers of the district – namely Payra, Bishkhali and Baleshwar – create an estuary which forms the beach covering an area of about four kilometres. This is a newly formed beach inaugurated for the tourists in 2017. At the entrance of the beach there is a huge *Jhao* (*Casuarina equisetifolia*) plantation which welcomes the tourists. The sunset is clearly visible from the beach, hence the name of the beach evolved as *Shuvo Sondha* (i.e., good evening). Lot of birds along with other wildlife are noticed there. There is a full moon festival at the beach when thousands of tourists from different districts gather there. This festival is organized by the district administration of Barguna to represent the beauty and potential of this beach. Tourists can also visit Ashar island from this area where many fishermen live and where fish are dried and processed. There is a Rakhine village at Taltali near Ashar island situated on the shores of the Bay of Bengal. Tourists have suggested improving the communication, accommodation, safety and security system of this area.

Wangala festival

The Garos is another ethnic tribe in Bangladesh mainly settled in the greater Mymensingh division. Wangala is the primeval tradition, thanksgiving and major religious festival of the Garo ethnic tribe (a matrilineal tribe). These tribal people used to offer their newly harvested crops to their gods, especially to *Missy* and *Saljong* known as their main gods of this religious festival. After dedicating their harvested crops to their gods they take their crops to their homes. Moreover, these people keep themselves busy with worship, dances and songs throughout the colourful festival day which occurs in the month of *Kartik* of the Bengali calendar (November–December). They wear their cultural dresses at the festival and dance with the Garo music. Wangala is not only a thanksgiving ceremony to their gods but also noteworthy as it embodies the unity and solidarity among the Garos. Residents from neighbouring villages and cities used to visit the community during the festival. The similar type of festival celebrated by the mainland people is known as *Nabanna* festival which is more festive.

Shiv Chaturdashi fair

Shiv Chaturdashi fair is an ancient Hindu religious festival which is known as one of the oldest and major cultural and religious fairs in the Indian sub-continent. This fair takes place at Shiva temple, one of the most historical and holiest places of the Hindu community, which is around 1,000 years old and situated at the top of the Chandranath hill (350 metres high from the mean sea level) of Sitakunda under Chottogram district. A four-day-long fair is annually organized during the *Chaturdashi Tithi* (14th night) of *Falgun* of the Bengali calendar (first week of March) when thousands of Hindu pilgrims, disciples and tourists gather at this temple for the worship of Lord Shiva expecting his holy blessings.

Other existing and potential festivals

Besides the above-mentioned major events and festivals at the nature-based tourism destinations, there are some more festivals which also take place at natural settings but these are more locally arranged where usually the local visitors used to visit these destinations. Some of these festivals are arranged for a very short period of time (perhaps only a few hours). Some of the examples of such festivals are *Polo* (bamboo fish traps), *Bawa* festival (300–400-year-old fishing festival taking place every two years in the marshy lands of Sylhet region), River tourism festival, Sun festival (an annual event organized at various venues to watch the first sunrise of the Gregorian calendar), *Jitiya/Mahale* festival (a social festival organized by Mahale ethnic tribe at Godagari of Rajshahi district), among others. Besides these regular festivals there are some other festivals like Kuakata beach carnival; Kite festival at St Martins; Tourism festival at Joyrampur, Jessore; Shakrain Kite festival; Pitha Utshob (cake festival); others are organized occasionally at various nature-based tourism destinations of Bangladesh.

Besides the above-mentioned nature-based events there are some potential nature-based events in Bangladesh. Proper planning, implementation, coordination, integrated monitoring and evaluation along with sufficient funding may develop and promote the following potential nature-based events/festivals in Bangladesh. Some of the examples are: fruit-based festivals like mango/litchi/jackfruit festival in Rajshahi, Chapai Nawabganj, Naogaon, Dinajpur, Satkhira, and so on; floating guava festival at Bhimruli over the Kritipasha canal in Jhalakathi and Barisal; plum, dragon fruit, watermelon, and so on, festivals; water lily festival at different destinations; Hilsha festivals at Barisal and Chandpur; other fish festivals; silk festival at Rajshahi and Chapai Nawabganj; honey-collecting festival; forest-based festivals like Sundarbans festival (Sundarbans Day: 14 February) at Khulna, Bagerhat and Satkhira; among many others.

ICT-based Marketing of Nature-based Event Tourism in Bangladesh

Promotion of these events as well as tourism destinations of Bangladesh are generally carried out through conventional marketing channels which warrant innovation (Ahmmed, 2013; Nekmahmud et al., 2021). However, the advent of various digital platforms and Internet facilities during the recent past has transformed the traditional marketing strategies to digital marketing strategies. Nowadays, various tour operators and travel agents of Bangladesh use social media (mostly via Facebook page/group or individual account) for marketing their services and offers. They like to promote their services with various promotional and discounts published in social media. They also like to send various offers as SMS, email and even direct phone call to the potential customers. Tourists also like to post their tourism planning, activities, experiences, complaints, feedback and reviews in their/service provider's social media which also promote the destinations and special events. Such tendencies of marketing (from both demand and supply sides) are growing day by day in Bangladesh which indicates a good potential of the application of ICT in tourism destination marketing in Bangladesh (see Figure 2.1).

The incorporation of ICT into the organizational framework of the destination marketing organization (DMO) is treated as a crucial factor to success (Vasiliki and Maria, 2013). Tourism stakeholders constantly look for the next step in the use and application of ICT to secure their businesses and to compete with their competitors (Ryan, 2011). Nowadays, almost every tour operator and travel agent of Bangladesh uses ICT services for conducting their businesses. However, still they are lacking innovative management models for providing the best services to their customers. Though there are several institutional frameworks and governance arrangements to guide, operate and manage tourism businesses in Bangladesh, there are acute problems in implementing those frameworks and arrangements in the field levels which jeopardize the destinations.

The recent data shows that there are officially 170 million mobile connections in Bangladesh. However, 90 million are unique subscribers (54%) as per the

Tour guides	GDS (Global Distribution System)		PMS (Property Management System)
Payment	Patrolling and protection	Photography via online platform	Availing tourism services
Branding	Issuing pass/permits	Promotion of tourism	Sharing information
Customer relation			Awareness development
Online ordering/booking	**Roles of ICT in event tourism**		Employment opportunity
Reporting/feed back			Internet and social media
Tracking progress	Other info like political, socio-cultural, environmental and institutional		Access up-to-date information
Online travel agents	Effective and efficient advertisement		Do's and don'ts about a destination
Security system	Products and service selection		DMS (Destination Management System)

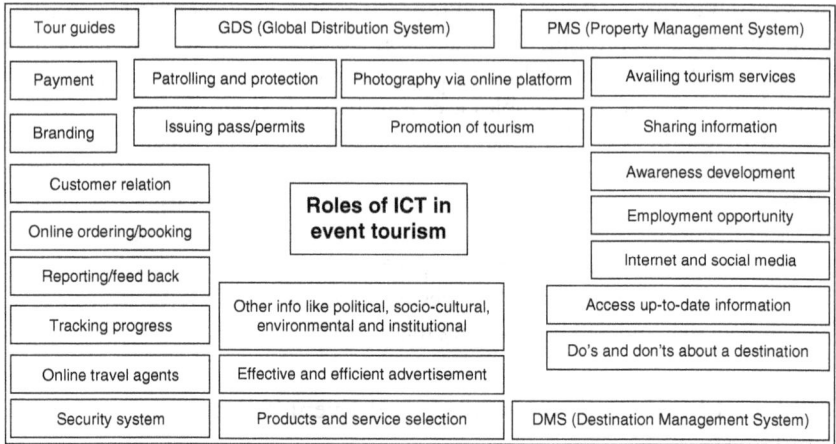

Figure 2.1 Roles of ICT in event tourism
Source: Author's compilation

data of December 2020. In Bangladesh, 41% of total mobile phone users have smartphones which is lesser than several south Asian countries like India (69%), followed by Sri Lanka (60%), Nepal (53%) and Pakistan (51%). Only 28% of the total population have the access of 4G connections though 95% mobile network in Bangladesh is under 4G coverage which is mainly due to the lack of affordability of devices, low levels of required knowledge and digital skills, safety and security issues. However, the 4G coverage in Bangladesh is better than Nepal (17%) and Sri Lanka (18%) (Staff Correspondent, 2021).

According to a recent report 'Digital 2021 Report Bangladesh', the total number of Internet users in Bangladesh in 2021 was 47.61 million (28.8% of total population with 19.2% growth) and active social media users was 45 million (27.2% of total population with 25% growth) where around 99% of them access social media through their mobile devices. There are 41 million Facebook users in Bangladesh (Kemp, 2021) which ranks first (84.3%) among the social media users followed by Twitter (9.6%), YouTube (3.21%), Pinterest (1.24%), LinkedIn (0.75%) and Instagram users (0.59%) (GlobalStats, 2021). Internet and digital technology can significantly contribute in driving economic growth and development in Bangladesh which can help to achieve the objectives of 'Perspective Plan 2021–2041', 'Sustainable Development Goals', and recover from economic crises due to the COVID-19 pandemic (Staff Correspondent, 2021), including the tourism and hospitality sector.

Recently, some advanced technologies have been applied in the Sundarbans mangrove forest for its better protection and management. As an example, a drone has been introduced in the Sundarbans mainly to monitor the forest against illicit felling and poaching which will also influence tourism to ensure security of the tourists during any event, particularly Rashmela and all year round. It will also be used for monitoring tigers and tiger habitats along with other major wildlife. Spatial Monitoring and Reporting Tool (SMART) and Global Positioning System (GPS) are being applied in the Sundarbans for an improved patrolling system, particularly for protecting the natural resources of the forest. However, there is limited mobile network inside the Sundarbans, particularly at the central and south zone of the forest. Nonetheless, the Forest Department has its own wireless telecommunication system to maintain its communications as well as protect the Sundarbans. It is expected to use some other technologies for taking better care of the Sundarbans as well as Rashmela festival to reduce its negative impacts.

Concluding Remarks, Policy Directions and Future Research

The study aimed to investigate the event tourism at the nature-based tourism destinations of Bangladesh and the contributions of ICT in their marketing, mainly promotion of these events. The study discussed seven such site-specific major events along with many other events and festivals which take place at the nature-based destinations of Bangladesh. Moreover, it discussed holiday-based events along with some other existing festivals and some potential nature-based festivals/events of Bangladesh.

Tourism marketing is part and parcel for promoting a destination either rural/ remote or urban in the map of global tourism market where ICT contributes significantly (Stratigea et al., 2008). Tourism is basically an information-intensive industry which warrants the applications of ICT. Nowadays, tourism and allied industries are largely dependent on the applications of ICT (Shanker, 2008; Zhang and Chen, 2019), which has created a new economy (Stratigea et al., 2008). Improved ICT service extension to event managers and event service providers can effectively and efficiently improve the marketing of event tourism through providing open and accessible information for the potential tourists which can facilitate event management, flow of revenues, natural and cultural resource management and conservation/preservation along with addressing and solving various challenges which ultimately foster the practice of sustainable tourism.

The modern-day tourists are more empowered now through access to the World Wide Web and a store of information. The tourists have more choices when they buy travel products and services due to available choices/alternatives/ options which are offered by online travel agents or tour operators and travel agents (Vasiliki and Maria, 2013). These tourists also expect personalized tourism services to enrich their tourism experiences which are categorized as before trip, during trip and after trip (Buhalis and Amaranggana, 2015). There are some advanced ICT tools like artificial intelligence (AI), augmented reality (AR) and virtual reality (VR), various computer and mobile device-based software/apps are used in hospitality and tourism sector (Nayyar et al., 2018), which might be used in promoting the nature-based events as well as other tourism products and services of Bangladesh.

Use of advanced ICT needs to be ensured at both enterprise (micro) level (like a small boutique family hotel/restaurant) and state (macro) level (like tourism ministry) to facilitate the growth of event tourism industry. It will also increase the competitiveness of tourism enterprises and organizations (Abdullayeva, 2019; Bethapudi, 2013; Zhang and Chen, 2019). However, there are several negative impacts of events. Commodification and overcrowding is one of them. Therefore, development of a niche market is essential particularly for religious tourism, nature-based tourism and ecotourism to reduce the negative impacts generated by disrespectful and environmentally unfriendly tourists (Vasiliki and Maria, 2013).

Bangladesh Tourism Board and Bangladesh Parjatan Corporation should pay special attention to upgrade ICT-based tourism marketing status of the country to fulfil the demands of modern-day tourists of both domestic and international origins. Concerned policies, strategies and frameworks should be revised to accommodate every marketing opportunity through addressing these key elements of ICT-based tourism marketing as well as stages of travelling to put forward the tourism sector of Bangladesh. Different initiatives to reduce digital divide are warranted. Rural development policies should address the issue to promote event tourism and to facilitate the overall welfare of nature-based destinations.

The level of environmental awareness and education of the people of Bangladesh is very low which is one of the main reasons for not managing the tourism destinations with due consideration. Uniqueness and authenticity of a festival warrants to be preserved by ensuring sustainability of the event which may be obtained through a mechanism of maintaining the authenticity of the rituals, educating and growing awareness of the visitors about their behaviour, activities and donations to help ensure that rituals can continue. Moreover, nature-based destinations require diverse marketing strategies including the applications of widespread ICT and marketing channels, hence they can focus on these issues (Vasiliki and Maria, 2013).

The establishments of alliances, partnerships, collaboration and networks among the tourism and allied stakeholders are considered as the most efficient relationships in tourism development. Public–private partnership (PPP) is recognized as an important strategy of building collaboration and leveraging the required resources for conducting tourism businesses successfully. This can also facilitate in building trust and building good relationships among the concerned actors (like tour operators, travel agents, service providers and tour guides) to ensure that tourism destinations can flourish. Such relationships facilitate stakeholders to learn from each other and leverage their scarce resources. It also assists them to collaborate together and make network (even virtual network) and sharing their online experiences to create new knowledge (Vasiliki and Maria, 2013).

There are very few studies on tourism marketing which impede desired tourism development and promotion in Bangladesh (Ahmmed, 2013; Ali and Mohsin, 2008; Karim, 2018). More scientific research activities are required on how to promote nature-based events with the application of ICT that enables the destinations to foster sustainable development initiatives. The effects of the COVID-19 pandemic on these events have been recognized as an emerging issue of academic discussions and research agenda. Moreover, how ICT can be applied to facilitate the recovery process of ongoing pandemics and how it can enhance the popularity of these events are some of the issues to be researched.

References

Abdullayeva, N. (2019). The role of ICT in tourism globalisation. *Paper presented at the 37th International Scientific Conference on Economic and Social Development – Socio Economic Problems of Sustainable Development.* Baku, 14–15 February 2019.

Ahmmed, M.M. (2013). An analysis on tourism marketing in Bangladesh. *International Proceedings of Economics Development and Research*, 67, 35–39.

Ali, M.M., and Mohsin, C.S.-e. (2008). Different aspects of tourism marketing strategies with special reference to Bangladesh: An analysis. *Business Review: A Journal of Business Administration*, 6(1), 1–3.

Aramendia-Muneta, M.E., and Ollo-López, A. (2013). ICT Impact on tourism industry. *International Journal of Management Cases*, 15(2), 87–98.

Backman, K.F., Backman, S.J., Uysal, M., and Sunshine, K.M. (1995). Event tourism: An examination of motivations and activities. *Festival Management and Event Tourism*, 3(1), 15–24.

Bethapudi, A. (2013). The role of ICT in tourism industry. *Journal of Applied Economics and Business*, 1(4), 67–79.

Buhalis, D., and Amaranggana, A. (2015). Smart tourism destinations enhancing tourism experience through personalisation of services. In I. Tussyadiah and A. Inversini (Eds.), *Information and communication technologies in tourism 2015.* Cham: Springer, pp. 377–389.

COMCEC (2015). *Effective tourism marketing strategies: ICT-based solutions for the OIC member countries.* Retrieved from: www.comcec.org/en/wp-content/uploads/2016/05/6-Tourism-Proceed.pdf (accessed 1 October 2021).

Dwyer, L., and Wickens, E. (2011). Event tourism and cultural tourism issues and debates: an introduction. *Journal of Hospitality Marketing and Management*, 20(3–4), 239–245.

Farkhondehzadeh, A., Karim, M.R.R., Roshanfekr, M., Azizi, J., and Hatami, F.L. (2013). E-Tourism: the role of ICT in tourism industry. *European Online Journal of Natural and Social Sciences*, 2(3s), 566–573.

Getz, D. (2008). Event tourism: Definition, evolution, and research. *Tourism Management*, 29(3), 403–428.

Getz, D., and Page, S.J. (2016). Progress and prospects for event tourism research. *Tourism Management*, 52, 593–631.

GlobalStats (2021). *Social Media Stats Bangladesh (Sept 2020–Sept 2021).* Retrieved from: https://gs.statcounter.com/social-media-stats/all/bangladesh (accessed 1 August 2021).

Gössling, S. (2020). Technology, ICT and tourism: From big data to the big picture. *Journal of Sustainable Tourism*, 29(5), 849–858.

Gössling, S. (2021). Tourism, technology and ICT: A critical review of affordances and concessions. *Journal of Sustainable Tourism*, 29(5), 733–750.

Januszewska, M., Jaremen, D., and Nawrocka, E. (2015). The effects of the use of ICT by tourism enterprises. *Zeszyty Naukowe Uniwersytetu Szczecińskiego. Service Management*, 16, 65–73.

Karim, Z. (2018). The impact of social media on tourism industry growth in Bangladesh. *International Journal of Economics, Commerce and Management*, 6(8), 463–482.

Kemp, S. (2021). Digital 2021 Report Bangladesh. *DATARPORTAL*. Retrieved from: https://datareportal.com/reports/digital-2021-bangladesh (accessed 1 August 2021).

Khan, Y., and Hossain, A. (2018). The effect of ICT application on the tourism and hospitality industries in London. *SocioEconomic Challenges*, 2(4), 60–68.

Nayyar, A., Mahapatra, B., Le, D., and Suseendran, G. (2018). Virtual Reality (VR) and Augmented Reality (AR) technologies for tourism and hospitality industry. *International Journal of Engineering and Technology*, 7(21), 156–160.

Nekmahmud, M., Farkas, M.F., and Hassan, A. (2021). Tourism marketing in Bangladesh. In A. Hassan (Ed.), *Tourism Marketing in Bangladesh*. Abingdon: Routledge, pp. 11–27.

Patwary, A.K., Chowdury, M.M., Mohamed, A.E., and Azim, M.S. (2020). Dissemination of Information and Communication Technology (ICT) in tourism industry: Pros and cons. *International Journal of Multidisciplinary Sciences and Advanced Technology*, 1(8), 36–42.

Ryan, V. (2011). Why information communication technology is a vital marketing tool with reference to the DVD/CD industry. *Technology Student*. Retrieved from: https://technologystudent.com/prddes1/ict1.html (accessed 12 September 2021).

Shanker, D. (2008). ICT and Tourism: challenges and opportunities. *International Journal of Engineering and Computer Science*, 5(4), 16293–16301.

Staff Correspondent (2021). Bangladesh behind Nepal, Pakistan in smartphone use says GSMA report. *The Daily Star*. Retrieved from: www.thedailystar.net/backpage/news/bangladesh-behind-nepal-pakistan-smartphone-use-2069457 (accessed 1 October 2021).

Stratigea, A., and Giaoutzi, M. (2017). ICTs and local tourist development in peripheral regions. In M. Giaoutzi and P. Nijkamp (Eds.), *Tourism and Regional Development*. Abingdon: Routledge, pp. 83–98.

Stratigea, A., Papakonstantinou, D., and Giaoutzi, M. (2008). ICTs and tourism marketing for regional development. In H. Coccossis and Y. Psycharis (Eds.), *Regional Analysis and Policy*. Cham: Springer, pp. 315–333.

Tahayori, H., and Moharrer, M. (2006). E-tourism: The role of ICT in tourism industry, innovations and challenges. *Journal of Applied Economics and Business*, 1(4), 67–79.

van Eck, G. (2018). The role of events on tourism. *Biz Community*. Retrieved from: www.bizcommunity.com/Article/1/595/185051.html (accessed 13 August 2021).

Vasiliki, K., and Maria, V. (2013). ICTs Integration into Destination Marketing Organizations (DMOs) tourism strategy. *Journal of Tourism Research*, 6, 67–77.

Zhang, K.-S., and Chen, C.-M. (2019). The impact of ICT on tourism business model: take ctrip group marketing as an example. *Paper presented at the 2019 IEEE Eurasia Conference on IOT, Communication and Engineering (IEEE ECICE 2019)*. Yunlin, 3–6 October 2019.

3   Exploring the Light Show Landscaping at Traditional Festivals and Events in China

Yuanyuan Zong and Muhammad Abdul Kamal

Introduction

Light show technology is a landscaping technology for an urban night tour and festivals and events, widely used across tourism cities worldwide. The outbreak of COVID-19 and resilience uncertainty caused the recession of outbound tourism in almost every country. Hence, the governments began to reopen domestic travel based on the status quo of epidemic control. In the post-COVID-19 era, China and its provincial governments have been emphasizing the night tour and related night festivals combined with culture creative tourism as promotional programs, which provided insights into the role of light show technology in landscaping for night tours, festivals and events of urban tourism. A recent focus from academia has been on the interplay of the light show, night landscape and tourism (Eldridge and Smith, 2019; Huang and Wang, 2018; Veronica et al., 2020). Some researchers have further discussed the principles of night landscaping with the application of lightscapes in cultural heritage (Valetti et al., 2020). To date, a plethora of studies have focused on the experience components of night tourism from a destination perspective, whereas tourist experience research on light show landscaping has been mostly ignored. To bridge the gap, this chapter undertakes a literature review followed by an interview with participants from four distinct regions. The content of the interviews then was taken to a thematic analysis. The findings are presented in three dimensions: light show landscape features; factors affecting light show landscaping; and tourists' emotional responses to light shows. Future research will also look into the effects of light show landscaping on destination festival marketing.

Literature review

Night-time tourism

Night-time tourism is defined as an extension of traditional day-time tourism activities into the night. Many cities have been touristified as the nocturnal destination. Museums, galleries and other attractions are required to stay open at night in accord with the metropolis policy, which contributes to the demand of

DOI: 10.4324/9781003271147-6

night tours (Eldridge and Smith, 2019). The conventional distinction between day and night has been eroded as daytime tourism has expanded into the night, providing tourists with a new experience (Markwell, 2018). The night tours have been developed into various forms like nocturnal zoo tourism (Markwell, 2018), museum night (Choi et al., 2020), astro tourism (Soleimani et al., 2019), ghost tourism (Dancausa et al., 2020), and so on.

Night tourism's attributes and cognition are described in a more open and transgressive manner, emphasizing the distinction between day and night. This conversation has transformed the night into a world of pleasures, excitement and thrill (Eldridge and Smith, 2019). Various tourists from various countries had different perceptions of the cityscape during the day and at night. The findings suggest that the night has brought out more personality as a result of the night view being connected with leisure, amusement and glamour (Huang and Wang, 2018). In another case on cultural heritage night, a multi-city night tour has been utilized to investigate how the notion of love marks is used as an emotional and cognitive response to night tourism experiences (Chen et al., 2020). The two dimensions of night tourism are the environment and the night atmosphere. Night lighting, as an aspect of the night ambiance, not only provides a sense of security and comfort, but it may also add to the aesthetic appeal (Veronica et al., 2020).

Lightscape in a tourism context

In terms of physical response to light, light expresses additional information in a peripheral and peaceful manner and stimulates to build atmosphere, provoke moods, and deliver immersive experiences (Yu et al., 2018). Casciani (2020) confirmed that human socially-oriented lighting standards are characterized by five characteristics: functionality; well-being; ambiance and urban beautification; relationship/experience; and environment. Based on these five features of lighting, light show technology has been widely used in the operation of tourism sites such as cities, attractions and cultural heritage. Lightscape has been built as a night tour experience object. Aside from tourists participating in nightlife, another aspect of a night trip is the spectacular night illuminated by lights (Shaw, 2018), which indicates the vital role of lighting in night tour sights. According to Eldridge and Smith (2019), lighting festivals and events produce an aesthetic effect by brightening the urban space and making it more lively. When tourists experience a location created by light show technology, they might gain a sense of place attachment and belonging. The images of light scaping can be identified using criteria such as "visibility", "standing out", and "relation with the settlement". The approach to compare the night view with daytime counterpart and analysis of lighting distribution form the design guidelines for the nocturnal image of the cultural landscape (Valetti et al., 2020).

A plethora of studies have focused on the tourist experience in the context of a destination's night tour. Scant academic studies have linked light show landscaping to tourism events and night tours. In the context of urban tourism and festivals, there is still a void in terms of investigating tourism elements, relevant factors and emotions of light shows. The integration of light show technology with city tourism and festivals is expected to be a promising area to dig into in the near future.

Light show technology applied in urban cities and traditional festivals

Light show technology is defined as a performing technology combined with lighting, audio, video, water fogging and electronic fireworks based on the attributes of destination and cultural connotations. It can generate a techno-logical and artistic performing space integrated with physical scapes in a des-tination (Zhihu.com, 2021). Light show technology is an agglomeration of technology and culture. In the level of technology, Chinese LST suppliers have offered the main three landscaping project technology: building projecting show; 3D Print Work; and front-projected holographic display (Zhihu.com, 2019). More updated yet assisted technologies include water projection, 5D projection mapping, drone light show, and so on. Drone light shows are the most advanced design in light show technology so far. A great number of illuminated, synchronized, and choreographed groups of drones arrange them-selves into various aerial formations. Almost any image can be recreated in the sky by a computer program that turns graphics into flight commands and communicates them to the drones. Drone light shows are free of framing in physical buildings and give the designers more creative space to lighten their imaginary world. In another level of culture, light show technology should be an effective communication language to fabricate the romantic story for festival and destination cities. More atmospheres are visualized and enhanced through light show technology.

In recent years, Chinese tourism cities have practised light show for festivals and events. The light shows have been popular night tour attractions for city tourists and are representative of four modes: CBD building light show, tourism sites light show, public area light show, and special-interest light show. Table 3.1 shows a sample of photos taken with the assistance of participants' contributions.

Methodology

Qualitative content analysis is one of the numerous research methods to ana-lyse textual data. This research method is used for the subjective interpret-ation of the content data through the systematic classification process of coding and identifying themes or patterns (Hsieh and Shannon, 2005). Being one of qualitative content analysis, thematic analysis has three stages of processing the data: abstracting the units of meaning, initial coding, and categorization (Braun and Clarke, 2006). In addition, thematic analysis has been vastly applied in the field of psychology, health care, and beyond (Xu and Zammit, 2020) as well as in the exploration of emerging phenomena like the relationship of tour group

Table 3.1 Modes and examples of light show in urban tourism city

Mode	Sample location	Sample photos
CBD building light show	Futian CBD, Shenzhen	
Tourism sites light show	Forbidden City, Beijing	
Public area light show	People's Square, Shanghai	
Special interest light show	multi-place drone light show in birthplaces of Communist Party of China	

members, hassles of aircrew, and other tourism issues (Tsaur et al., 2019; Tsaur et al., 2020). Therefore, this study adopts thematic analysis to identify the impressive elements of light show landscaping, the factors influencing light show landscaping, and tourists' emotional experience on the light show.

The qualitative data were sourced from in-depth interviews from July to August 2021. The interview participants are recruited online and contacted through

Table 3.2 Summary of participants

ID	Gender	Age	Country/region	Destination
T1	Female	23	China Mainland	Beijing, Shanghai, Shenzhen, Hefei
T2	Female	26	China Mainland	Shanghai, Hong Kong, Xi'an
T3	Female	28	China Mainland	Beijing, Shanghai, Wuhan
T4	Female	30	China Mainland	Hangzhou, Shanghai, Guangzhou, Chengdu,
T5	Female	32	China Mainland	Hong Kong, Beijing, Shanghai, Dalian
T6	Female	40	China Mainland	Shanghai, Xiamen, Beijing, Hangzhou
T7	Male	22	China Mainland	Beijing, Shanghai, Nanjing
T8	Male	25	China Mainland	Chengdu, Wuhan, Qingdao
T9	Male	27	China Mainland	Hangzhou, Beijing, Yantai
T10	Male	32	China Mainland	Shanghai, Guangzhou, Fuzhou,
T11	Male	35	China Mainland	Shanghai, Xi'an, Zhenzhou,
T12	Male	39	China Mainland	Beijing, Shanghai, Changsha, Urumqi
T13	Female	29	Taiwan	Beijing, Shanghai, Nanjing, Xiamen
T14	Female	33	Taiwan	Shanghai, Changsha, Chengdu
T15	Female	35	Taiwan	Beijing, Shanghai, Shenzhen
T16	Male	28	Taiwan	Shanghai, Hangzhou, Guangzhou
T17	Male	41	Hong Kong, China	Beijing, Shanghai, Xi'an
T18	Male	36	Hong Kong, China	Beijing, Shanghai, Wuhan,
T19	Female	27	Malaysia	Shanghai, Shenzhen, Suzhou
T20	Male	28	Malaysia	Beijing, Shanghai, Xi'an

personal social networking. Many tourists from China Mainland share their travel experience with pictures, photos and associated descriptions in interest groups online like Douban(豆瓣). Through Douban platform, likes and following posts are the common approaches to contact with like-minded travel friends. Twelve participants from China Mainland have recruited through the Douban interest group. The researcher contacted 5 participants from Taiwan, Hong Kong and Malaysia through personal recommendation. Finally, 20 participants have been selected, of which 12 participants were from China Mainland, 4 from Taiwan, 2 from Hong Kong, China, and 2 from Malaysia. In the last five years, all of the participants had visited Chinese cities to see light shows at festivals. Table 3.2 contains their demographic information. The researcher requested the participants to recall their impressive experience of seeing light shows at festivals and events while travelling in a city destination within one hour and then following questions have been asked from them:

Q1: What aspects of light show landscaping at festivals and events made you feel impressed over your five years of traveling to China's tourism cities?

Q2: What factors have influenced light show landscaping?

Q3: Please elaborate on the various emotions that these light shows have evoked in tourists.

Table 3.3 Thematic analysis of interview textual content

Themes/Category	Frequencies
Light show landscaping elements	188
1.1 Tourism resources	43
1.2 Aesthetic image created by light show technology	38
1.3 Traditional festival and events symbols	32
1.4 Advertisements	30
1.5 Slogans for Political events	26
1.6 Background music	19
Factors influencing light show landscaping	153
2.1 Marketing strategy of night tour in city	45
2.2 Traditional festival culture and ceremony	40
2.3 Chinese popular culture and social values	36
2.4 Propaganda policy for political power and national growth	32
3. Tourists' emotional experience on light show	48
3.1 Delightful	20
3.2 Inspiring	16
3.3 Nostalgic	12

The destination cities referred by participants were the representative metropolitan cities like Beijing, Shanghai, Hong Kong, Xi'an, Shenzhen, Hangzhou, Nanjing, Chengdu, and so on. The theoretical saturation has been reached when the 18th participant completed the interview (Robinson, 2014). All the transcripts of the interview have been recorded, double-checked and transformed into textual data. In the process of categorization, the author and another expert teamed up as coders to independently review the data and code them into meaningful units. In total, 410 meaning units were categorized and the 389 valid units remained with the interrater reliability rate of 95% (389/410). Based on differences and similarities, the sub-categories and categories were sorted from various codes through the process of reflection and discussion between two coders. The latent content of the categories was further formulated into a theme in the agreement of categorization results (Graneheim and Lundman, 2004). Finally, the researchers summarized six categories based on Q1 data. Four categories and three categories were sourced from thematic analysis on the transcripts of Q2 and Q3. The frequency of these categories is shown in Table 3.3. Moreover, the interrater reliability rate (IRA) for each category was above 0.88, indicating a significant result of categorization (Kassarjian, 1977).

Results

The results of the in-depth interviews revealed three themes and 14 categories based on the tourist experience of watching light shows at festivals and events. The tourist experience influenced by light show technology was divided into three themes: light show landscaping elements; factors influencing light show landscaping; and tourists' emotional reactions to light shows.

Light show landscaping elements

Tourism resources

Based on the frequency in descending order, light show landscaping elements are categorized as "Tourism resources", "Aesthetic image created by light show technology", "Traditional festival and events symbols", "Advertisements", "Slogans for Political events" and "Background music". The light show stages the tourism elements most, such as landmarks of natural landscapes, buildings, symbolic flora and fauna, history and culture. For example, some participants stated:

"Light show at festivals held in Xi'an are often framed around the skyline of buildings with Ancient capital Chang'an style, creating a bustling night view of Chang'an in Tang Dynasty" (T11).

"Light show is an effective approach to get to understand local tourism resources. Wuhan Sakura festival 2021 presented a light show themed at the cherry blossom tour. There are many fabulous views of cherry blossom in Wuhan University and East Lake as famous attractions for tourists" (T3).

Aesthetic images created by light show technology

With the application of light show technology, lighting and shades generated aesthetic images by adjusting the brightness of light, colour tune, and transform-ations of patterns. Furthermore, with the help of drone light show technology, exquisite and dynamic aesthetic landscaping such as a starry sky, sea of flowers, waves of the sea, snow falling, and abstractionism figures have been produced. One participant has remarked:

"The traditional Chinese festivals have been related to the world of tales and legends. The backgrounds of illusion and dynamics like flowing clouds, the eclipse of the moon, blossoming flowers turn to glorious views for festivals and events" (T8).

Traditional festival and event symbols

Festivals and events are the timing for light show performances in metropolitan cities. The symbols of traditional festivals and events are ritualized in tales, and so produced as essential aspects of light shows. For example, the light displays on Chinese New Year's Eve contain a couplet of blessings, tree sprouts, new year's greetings, the Chinese Zodiac animal for the following year, and so on. For instance, one participant stated: "The related cultural emblems are displayed in light shows during traditional festivals and events. After all, the festivals are linked to a unified Chinese culture over the world. The moon and Moon Goddess Chang'e's white rabbit, for example, are obviously part of the Mid-Autumn Festival light show design" (T13).

Advertisements

Light show functioned as an advertisement in city festivals and events. Several light shows are financed by firms or advocated by the local government. Some participants stated:

"Advertisements are cut into a short section when watching light shows. Commercial and public service announcements are included in various aspects of the light show. Over the last three years, the most common public service announcements (PSAs) have been related to anti-COVID-19, and PSAs like washing your hands regularly occur frequently in light shows" (T1).

Slogans for political events

As a means of interacting with the public, light shows may include political slogans, which can be a unique feature of light show landscaping. To a large extent, light show technology is used to propagate slogans at the opening of the Chinese National Congress, national policies and strategies, and technological achievement events. Participants have remarked:

"When it comes to the important political ceremonies and events, the emblem and flag of the CPC, red star, group portrait of working people and Chairman Mao are symbolized slogans in the light show" (T7).

"The light shows at Urumqi are centred on celebrating the building achievements of Silk Road, which conclude the symbolized slogans like united multi-ethnic cooperation" (T12).

Background music

The way of experiencing light shows is not merely through visual sense, but also through auditory sense. The landscaping of the light show can attain a better effect by multi-sensual integration. The background music is selected and choreographed in line with the topic of the light show, visual images, audience preference, and popular trends. In participants' comments, the background music is an intangible element of light show landscaping and bring their different experience when appreciating the light show. The types of background music chosen for the light show are traditional festival songs, western classic symphony, New Age-styled songs, the original soundtrack of films, and popular songs.

"Light shows on National Day select classic patriotic songs as the background music. The well-known lyrics and melodies combined with changes of light show landscaping, expressing the magnificence of Chinese rivers and mountains" (T14).

Factors influencing light show landscaping

Four factors influencing light show landscaping are summarized as marketing strategy of night tour in the city, traditional festival culture and ceremony, Chinese pop culture and social values, and propaganda policy for political power and national growth. Due to the impact of COVID-19, Chinese cities stick to the policy of the tourism bubble and the diversion of tourist flows. Thus, night tour in urban festivals has been promoted by Ministry of Culture and Tourism in China recently. Light show landscaping is utilized for night tour marketing alongside urban space beautification and safety function. For instance, one participant stated:

"Recently, almost every city implemented light show as a channel for tourism festival marketing. The light show landscaping is determined by local night tour resources. The hype package of Hangzhou tour is designed as two must-see tours: West Lake tour at daytime, Light show at Qianjiang at night" (T6).

The light show is relevant to Chinese traditional festivals and ceremonies. The Chinese Lantern Festival, which has been held on the 15th of the first lunar month since the Tang Dynasty, is inspired from a Buddhist rite of lighting a candle for prayer. The Chinese people spread light show practices to all kinds of traditional festivals and shaped the aspects of light show landscaping by carrying on the festival ritual and culture. One of the typical comments is:

"What elements are expected to be staged in a festival light show must adhere to Chinese traditional festival culture and ceremonial. Although some light displays are technologically innovative, the content of light show landscaping is determined by the festival culture" (T19).

The selection of elements in a light display is influenced to some extent by Chinese popular culture and social norms. Early on, the light shows were of poor quality in terms of colour and choreography, and they resembled light show performance for disco dancing. Over 10 years, the light shows have gradually been reflecting pop culture and social values at festivals, with references to worldwide design coupled with Chinese characteristics. Blockbusters, Internet slang, and animation all reflect pop culture. Eco-friendly activities, social peace and gratitude for others' assistance were all examples of social values. Light show landscaping has been influenced by positive popular culture and social values.

"Drone light show in Wuhan Lockdown-lifting Day have staged a group portrait of doctors and nurses to show great gratitude toward medical team" (T18).

Furthermore, propaganda public policy for political power and national progress has influenced light show landscaping in China, which is also a technological and creative response to propaganda. "The propaganda policy is an influencer of the originality of light show landscape. For example, the light show should creatively storytelling the history of achievements in the commemoration of Shenzhen opening and reform. So the bronze Ox is designed as a symbol for the upsurge of city finance and stocks, while the chemical equations are patterns representing a growth of the bio-tech industry of Shenzhen" (T15).

Tourists' emotional experience on the light show

The tourists evoked three emotions when watching the light show at festivals: delightful, inspiring and nostalgic. The participants expressed that light shows can bring them delight visually at night festivals. A typical opinion is presented as: "The light shows in metropolitan cities at festivals are a drive to lighten the magic city night. The Drone light show and 5D light mapping are always making everyone delightful" (T16).

The inspiring emotion is sourced from the performance of light show technology and the elements it demonstrates. One participant remarked: "When seeing the slogans arranged by drones in skies, like how marvellous, my China (厲害了我的國), I feel inspired on the future development of science and technology" (T14). Several participants would be nostalgic for a bygone era owing to the light show elements on hometown and national buildings, complemented by background music.

"I wander among the CBD buildings where light shows were mapping in Shenzhen. There were lots of images on waves and sparkling stars to create the theme of national constructions. The slogan of 'I love China' was a cue to recall the nostalgia for the roots-seeking history of oversea Chinese" (T19).

Discussion and Conclusion

This study conducted an in-depth interview with 20 participants to analyse the tourist experience of light show landscaping at traditional festivals and events. Three themes were summarized as light show landscaping elements, factors influencing light show landscaping, and tourists' emotions regarding the light show. Tourism resources, the aesthetic image created by light show technology, traditional festival and events symbols, advertisements, slogans for political events and background music were six categories associated with the theme of light show landscaping elements. Factors influencing light show landscaping include marketing strategy of night tour in the city, traditional festival culture and ceremony, Chinese pop culture and social values, and propaganda policy for political power and national growth. The tourists will have three emotions when watching lights shows at festivals.

Although Eldridge and Smith (2019) have mentioned the light show in urban night tour have impacts on tourist experience, there is a notable knowledge gap with regard to tourist experience of the light show. Thus, this study attempted to fill the gap on interdisciplinary research of light show technology and tourism experience. Based on the fact that light shows at festivals in China serve as an advertisement tool and a platform of propaganda, the landscaping elements of the light show will be balanced between urban tourism positioning, festival travel image creation and urban functional planning. The findings will be a reference for light show landscaping in other Asian countries as well as a guideline on how to integrate light show technology into night tours and festivals. More warm-hearted storytelling and plot design in a light show need to be considered. Besides, the issue of eco-friendly light show landscaping will be a concern for the practical operation. The tourist emotions of watching the light show are also impacted by tourist cultural background. The future research can be focused on discussion on the consequences of the light show on tourist loyalty, revisit intention and place attachment. Historical nostalgia and national identity can also be considered as moderators or mediators in empirical study on the experience of the light show.

References

Braun, V., and Clarke, V. (2006). Using thematic analysis in psychology. *Qualitative Research in Psychology*, 3(2), 77–101.

Casciani, D. (2020). *The Human and Social Dimension of Urban Lightscapes.* Cham: Springer.

Chen, N., Wang, Y., Li, J., Wei, Y., and Yuan, Q. (2020). Examining structural relationships among night tourism experience, lovemarks, brand satisfaction, and brand loyalty on "Cultural Heritage Night" in South Korea. *Sustainability*, 12(17), 6723.

Choi, A., Berridge, G., and Kim, C. (2020). The Urban Museum as a sustainable tourism attraction: London Museum Lates Visitor Motivation. *Sustainability*, 12(22), 9382.

Dancausa, G., Hernández, R.D., and Pérez, L.M. (2020). Motivations and Constraints for the Ghost Tourism: A Case Study in Spain. *Leisure Sciences*, DOI: 10.1080/01490400.2020.1805655.

Eldridge, A., and Smith, A. (2019). Tourism and the night: Towards a broader understanding of nocturnal city destinations. *Journal of Policy Research in Tourism, Leisure and Events*, 11(3), 371–379.

Graneheim, U.H., and Lundman, B. (2004). Qualitative content analysis in nursing research: concepts, procedures and measures to achieve trustworthiness. *Nurse Education Today*, 24(2), 105–112.

Hsieh, H.F., and Shannon, S.E. (2005). Three approaches to qualitative content analysis. *Qualitative health research*, 15(9), 1277–1288.

Huang, W.J., and Wang, P. (2018). "All that's best of dark and bright": Day and night perceptions of Hong Kong cityscape. *Tourism Management*, 66, 274–286.

Kassarjian, H.H. (1977). Content analysis in consumer research. *Journal of Consumer Research*, 4(1), 8–18.

Markwell, K. (2018). An assessment of wildlife tourism prospects in Papua New Guinea. *Tourism Recreation Research*, 43(2), 250–263.

Robinson, O.C. (2014). Sampling in interview-based qualitative research: A theoretical and practical guide. *Qualitative Research in Psychology*, 11(1), 25–41.

Shaw, R. (2018). *The Nocturnal City.* Abingdon: Routledge.

Soleimani, S., Bruwer, J., Gross, M.J., and Lee, R. (2019). Astro-tourism conceptualisation as special-interest tourism (SIT) field: A phenomonological approach. *Current Issues in Tourism*, 22(18), 2299–2314.

Tsaur, S.H., Cheng, T.M., and Hong, C.Y. (2019). Exploring tour member misbehavior in group package tours. *Tourism Management*, 71, 34–43.

Tsaur, S.H., Hsu, F.S., and Kung, L.H. (2020). Hassles of cabin crew: An exploratory study. *Journal of Air Transport Management*, 85, 101812.

Valetti, L., Pellegrino, A., and Aghemo, C. (2020). Cultural landscape: Towards the design of a nocturnal lightscape. *Journal of Cultural Heritage*, 42, 181–190.

Veronica, S., Ginting, N., and Marisa, A. (2020). Local wisdom-based on development of the environment and atmosphere aspect of Berastagi Night Tourism. *International Journal of Architecture and Urbanism*, 4(2), 144–155.

Xu, W., and Zammit, K. (2020). Applying thematic analysis to education: A hybrid approach to interpreting data in practitioner research. *International Journal of Qualitative Methods*, 19, 1609406920918810.

Yu, B., Hu, J., Funk, M., and Feijs, L. (2018). DeLight: Biofeedback through ambient light for stress intervention and relaxation assistance. *Personal and Ubiquitous Computing*, 22(4), 787–805.

Zhihu.com (2021). *The main technology of light show in various tourist sites* (in Chinese). Retrieved from: https://zhuanlan.zhihu.com/p/391379471 (accessed 20 September 2021).

Zhihu.com (2019). *Experts from Zhongyi light show company discuss building projection, 3D mapping and front-projected holographic display* (in Chinese). Retrieved from: https://zhuanlan.zhihu.com/p/76363749 (accessed 20 September 2021).

4 Goddess in Digital Space
A Study on Dynamics of Digitalization in Autumn Festival of India

Samik Ray

Introduction

Festival is the culturally significant set of traditionally organized or newly developed collective pleasurable displays (Quinn, 2009; Janiskee, 1980) of communal identity (Matheson, 2005) created (Turner, 1982), staged and celebrated by a community and also a creative spend of social space and time for socialization (Ray, 2018) and periodical renewal of community life-stream (Falassi, 1987). Festival is possibly the earliest event form, thus a subset of the event and has a long historical trajectory (Quinn 2009). Since food plays a decisive role in survival, the celebration of food accumulation, particularly hunting, was crucial to the primitive society while harvesting celebrations to the agrarian. Hunting moved from a survival deed to leisure-adventure centric entertainment with the qualitative changes in food production technology and the emergence of an agrarian society. Festival-event rolls through the ages with qualitative changes in nature, activity, emotion, experiences of celebration applying contemporary technological innovations. It celebrates the religious ceremony, season change and harvesting, historic and heroic event commemoration, and artistic expressional and traditional folk performances. Festival transforms into a carnival when traditional ceremonies combine various tangible and intangible cultural displays, parades, public street parties and attract a considerable number of non-resident crowds and travellers.

Tracing the Festival-Event-Technology-Tourism Interaction

A shift from primitive rationale to community gathering motivation led the festival to open up a social space for socialization, socio-cultural bonding (Ray, 2018), and community identity display. The beginning of festival-centric entertainments and trading made the festival socio-economically significant, creating a substantial space for cross-community mobility within its periphery. The festival-event then began to define place identity. With the growth of cross-community and cross-boundary mobility to the festive locations in the late pre-Christian era, imperial merchant patronage and technology application became crucial to prepare the place with the required infrastructure, services and facilities to host an

DOI: 10.4324/9781003271147-7

orderly festival-event. Festival turned to a significant socioeconomic institution within medieval and colonial fabric since cross-boundary mobility of commoners, travellers, merchants, colonials, inquisitive, cultural performers, entertainers and producers across regions (Ray, 2018) with varied motivations began to attend the festivals. Interaction between festival-event and tourism has increased substantially with the diversification of attendance motives and activities. It entered into an era of reciprocal and mutual dependency in sheer support of nineteenth-century technological innovation-aided logistics for better visitor/tourist flow. Festivals emerged as a motivator of organized travel and tour packages after the post-1840 experiment of organizing event-centric package tours by Thomas Cook and his contemporary tour companies.

At the dawn of the twentieth century, festival-events and tourism began to co-evolve while sharing technology-supported similar logistic mechanisms, promoting the same destination, using the same attraction as motivation, and becoming the prime driver for place identity and attractiveness. Festival-event's role in interaction with tourism became different when it turned into a niche commodity in the tourism and event market with the post-1950 rapid growth in tourist footfall and mass tourism market. Festival-event, then, gradually changed from being driven by traditional socio-cultural emotion and significance to tourism market-driven losing traditional ethos of communal creative displays to staged authenticity. The twenty-first-century Web revolution redefines festival, technology, and tourism interaction, countering all the post-1950 approaches and aspects that emerged to turn festival-event into a mass tourism market-driven commodity, re-evaluating commodification; commercialization; authenticity; and community participation concepts in tourism.

Academic approaches

Interaction among travel and hospitality logistics, technology, and festival-event contributed to shaping and changing the perception and tradition of entertainment, creative socialization, and spending leisure time over the ages but did not receive wide academic attention until the 1990s. Among the early studies, Greenwood (1972) and Gunn (1979) noted the festival's significance in tourism while Janiskee (1980) investigated its socio-cultural and economic impact. Scotinform Limited's (1991) study on festival visitors and Getz's (1991) discussion on festivals, special events, and tourism were noteworthy among the early 1990s investigations. In the last three decades, researchers mostly carried the investigations and studies on visitors' motivation to attend the festival-events (Backman et al., 1995; Scott, 1996; Crompton and McKay, 1997), festival motivation (Mohr, 1993; Schneider and Backman, 1996; Lee et al., 2004), socio-cultural and economic impact (Crompton and McKay, 1994; Getz, 2000; Small and Edwards, 2003; Xie, 2003; Matheson, 2005; Chang, 2006; Picard and Robinson, 2006; Quinn, 2009; Phipps and Slater, 2010), festival-event's contributions to tourism (Getz, 2000; Quinn, 2009; Allen et al., 2010), the role of festival-events in destination image and identity building and reorganization of branding (De Bres and Davis, 2001; Prentice and Andersen, 2003; Mossberg and Getz, 2006; Snowball and Willis, 2006; Ferdinand and Williams, 2013; Shabnam et al., 2018), marketing and policy approaches (Formica and Uysal, 1996; Gotham, 2005; Quinn, 2010), resident perception (Cudny et al., 2012), and types (De Valck, 2007; Cudny, 2011; Cudny and Rouba, 2011) aspects and issues.

In the twenty-first century, scholars begin to state festivals as tourist assets (De Bres and Davis, 2001; Kowalczyk, 2001; Gotham, 2005; Markwell and Waitt, 2009; Cudny, 2011; Cudny et al., 2012; Markova and Boruta, 2012). Although recent quantitative growth of festival centric tourism led Mika (2007) and Cudny (2011) to consider it as festival tourism or a distinct variety of tourism, Buczkowska (2009) treats it as a form of cultural tourism, while Getz (2008, 2010) as a subset of event tourism only with the synergic merger of event and tourism (Ray, in press). Research on issues and impacts of technology application in tourism hospitality grew significantly with the growth of information and communication technology (ICT) application across society. Studies on ICT's application in event tourism mainly focus on promotion and marketing (Arora and Sharma, 2020; Djumrianti, 2018) and issues and impacts in tourist behaviour, organizational performance, and operation management (Breukel and Go, 2009; Vu et al., 2017, 2018). Topics related to interrelation and interaction among tourism, technology, and the festival-event do not receive academic attention yet.

Festival-centric tourism

Although festival-events gained the commodity identity in the mass tourism market but turned into niche leisure and wanderlust travel attractions with demand increase for specialized visitation experience that connects visitors with destination and hosts culturally in the late twentieth century. In the twenty-first century, festival-events becoming a tourism driver or motivator contribute to raising the attractiveness, positioning and competitiveness of tourism offers at the known tourism destinations and place the unknown or lesser-known destinations in the tourism map. Collaboration between festival-events and tourism increases with the logistics and promotional mechanism sharing by both. It escalates mutually dependent cross-sector activity that successively encourages cross-promotion or cross-sells, ensures regular tourist flow multiplying revenue, causes a makeover in the place economy, and builds a better destination image. All those phenomena together open up the growth opportunity of festival-centric tourism or festival tourism. It is becoming popular across the world. Munich Oktoberfest, Valencia's Las Fallas, Harbin's Ice Festival, Bhutan's Tsechu, Nagaland's Hornbill Festival, Guatemala's Semana Santa, Songkran of Thailand, Obon's festival of Lanterns, Perahera of Kandy, and carnivals of Italy; Notting Hill; Tobago are a few examples of festive-centric tourism. Durga festival of Kolkata has also made its place on the festival tourism map of late.

Durga Festival: Myth to Carnival

Hindus worship Bāsanti-Durga in Spring (March–April), the period of enlighten-ment and light when gods stay awake and Shārad-Durga in Autumn (September–October), the period of darkness when gods sleep and demons awake. According to Hindu lore, a proverbial Bengal king named Surath and a merchant called Samadhi Vaishya initiated the Bāsanti-Durga celebration in 4th–6th CE. Hindu tradition of victory celebration over evil spirits in 'Dussehra' (Pan India) and Shārad-Durga festival or 'Akalbodhan' (eastern and north-eastern India) has emerged from Rama lore. The lore depicts Lord Rama's untimely (*akāl*) effort to awake (*bodhan*) the goddess and then worship her for blessings to fight evils. The tradition of nine different newborn plants (*nabapatrikā*) worshipping during the festival originated from the harvesting ceremony. Durga image, worshipped across India, is not homogeneous (McDermott, 2001). Eastern India worships the 'Mahishāsuramardini' (Mahish-buffalo, asura- demon, mardin-female slayer) form linked to the Durga narrative of rinsing evils out from the world killing the notorious buffalo demon.

In the eleventh and twelfth centuries, only warriors and royals were worship-ping Durga, idealizing her as the epitome of strength, rage, power, kindness, and protection (Handiqui, 2011). By the sixteenth century, it turned into a festival of powerful, wealthy landed and merchant classes. Autumn festival, then, opens up opportunities for the wealthy class to showcase socio-economic grandeur, socialize exclusively among themselves, cross-generation gathering within their families, and married girls to visit the parental home with children. A shift from the practice of worshipping solo Durga to the Durga family idealizing the con-cept of her annual visit to the parental home on the earth with children emerged when the social narrative about married girls visiting parental home diffused into the festival.

The festival began to lose its class affiliation attached to the wealthy and powerful when a conflict with the non-wealthy began over mass participation and access to the celebration. According to local narratives, by the late eighteenth century at the rude denial of non-wealthy entry to the wealthy Sen family celebra-tion site at Guptipārā, near Calcutta, 12 (*bāro*) local impecunious priest friends (yāri) set a similar celebration in motion for their community. It developed as a '*bāro-yāri*' celebration with the spread of similar initiatives across Bengal among different non-wealthy communities. Although '*bāro-yāri*' celebration led the fes-tival to shift its affiliation from class to the community, the mass participation seemed to be impending there as in the festival context '*bāro-yāri*' then relates to the feeling of overt community centrism though originated as the opposite to the wealthy and powerful. By the early twentieth century, the Durga festival was no longer a preserve of powerful-wealthy families or particular communities as '*sarbajanin*' (for all the people) celebration emerged with opportunities for all the people to contribute to and participate in the festival and take its own-ership. As the '*sarbajanin*' celebration rises above all kinds of socio-economic divides, the post-1905 nationalist leaders used to utilize the occasion to promote indigenous products against colonial ones publishing advertisements in the vernacular newspapers and putting the sale counter at celebration sites.

In the post-1980s, '*sarbajanin*' celebrations spread across the east and northeast of India, while it expanded at a brisk pace in Kolkata. The state of West Bengal in 2019 identified 5,000 registered '*sarbajanin*' celebrations in Kolkata (Niyogi and Mukherji, 2019), and approximately 18,000 more were from registered housing societies and heritage houses. By the twenty-first century, the festival enters into the growth era while spreading across the city. The city turns into an ongoing exhibition of kaleidoscopic wonder transferring everyday space into an illuminated magical land of newly created or recreated architectural and sculptural marvels out of anything and everything within the frame of the temporary temple, its surround and idols employing 100,000 artists, artisans, craftsperson, architects, priests, drummers, painters, technicians, professional and amateur event organizers and many others for planning, designing, and execution. Thus each of the 28,000 festival locations sustains a micro-economy (Niyogi and Mukherji, 2019; Banerjee, 2019). They get into competition for titles or awards that foster the festival spirit. Opportunities to create a cultural bonding with the local market and capture the festival's emotional appeal of the locals and visitors lead corporate, multinational, and manufacturing companies to come forward with sponsorship and funding.

The economic worth of the creative activities around the festival was around £3,290 million when sponsorship from large multinationals and Indian corporate together was £32 million per capita retail sales and food-beverage were £288 million and 28.1 million in 2019 (British Council, 2019). The economic contribution of the festival to the state's GDP usually fluctuates between 3% and 10% (British Council, 2019; Banerjee, 2019). Corporate spends for advertisement was INR (Indian rupee) 6,50,00,00,000–9,50,00,00,000 in 2019 (Niyogi and Mukherji, 2019). Thus Durga festival gradually transforms from being driven by traditional socio-cultural emotion and significance to a tourism market-driven carnival of a magnificent and imposing socio-cultural and public art show combining tradition and modernity in various tangible and intangible cultural displays, parades, public street parties amidst millions of locals and visitors. Of late, the State government started an immersion carnival following the Rio Carnival. This festival now happens to be a crucial motivator and driver for Kolkata's tourism. It raises the attractiveness, positioning and competitiveness of tourism offers. Indeed, a sharp spike in tourist footfall substantiates it, such as 2019 record arrival of 1,487,688 tourists, 16.66% spike in air traffic, 10.67% spike in train traffic during Durga festival at the city (British Council, 2019).

Technology's Interference

The tradition of technology application probably began with idealizing the Goddess as a symbol of supreme military strength empowered by contemporary war-technology aided arms and ammunition. Indeed, it is always either socioeconomic status or better logistics requirement driven but remains instrumental to the Durga festival's quantitative growth and qualitative changes in logistics, pattern, format and cultural displays. The qualitative transformation in materials, style, and technique of construction or show occurred with civil-site construction technology advances as is evident in changes from a simple sage dwelling worship-site to a magnificently recreated glittering temporary classical or modern and postmodern structure. Likewise, shifts from bamboo to burnt clay brick or stone and then to matchsticks, pins, abandoned black disc records, and metals are also evident. In the growth era and then in the carnival era Durga festival gets facilitated by improved construction techniques, illumination mechanisms, logistics, and post-1950 print and audio-visual media innovation-aided publicity and operation. Worries about the festival's ill impacts on the city's surface and water bodies led all who work for the festival, switch to green practices using sustainability and carrying capacity concern technology. Thereby, organic and detoxification technologies are in use to prevent or control environmental pollution at the preparation phase. The use of hydraulic and water treatment and solid waste management technologies also become crucial to clean the post-festival city environment.

Dynamics of Carnival's Ride on Digital Technology and ICT

ICT and digital technology applications in Kolkata's autumn extravaganza favoured recent transformation in the festival and festival-centric tourism experience, marketing, imaging, publicity, and operation through a bottom-up trajectory. With the rise of the Internet and social media users respectively to more than 800 and 600 million, and the fast growth in smartphone use in the last 12 years, Indians across socio-economic spheres became digitally armed (Statista 2021a, 2021b). Of late, digital natives and residents (Prensky, 2001a, 2001b) turned to be the vanguard of transformation while initiating practices of sharing carnival information and impression in digital space. It makes the festival resurge with new socio-cultural dynamics amid changing celebration ecosystem, resulting from the interactions between the festival and digital technology. Today carnival experience is nowhere if visitors are digitally unarmed. Thus this practice recently spread rapidly, switching digital immigrants (Prensky, 2001a, 2001b) and non-resident temporarily to the resident category during the festival.

Carnival photos and videos uploaded and shared in digital space and social media by them certainly propagate and promote Kolkata and its carnival around the globe much faster and more effectively than before, making its worldwide real-time virtual presence possible. It prompts far-sighted and pragmatic festival-event organizers, tourism stakeholders, you-tubers, ICT professionals, and companies to understand the importance of ICT and digital technology application to the growth of competitiveness and then to offer virtual carnival tours, online offering and religious services, live streaming of rituals, cultural displays and performances, online purchase-cum-delivery of Goddess grace; gifts; offerings through Web. In 2015 these websites recorded more than 50,000 hits at carnival time (Ghosal,

2016). Since 2018 a large group of '*sarbajanin*' celebration organizers without developing their websites prefer to make a business tie with professionally managed IT companies for online live streaming of cultural and ritual displays, promotion, marketing, and celebration related e-services. Those phenomena, spreading with greater mobility in recent years, make Bengali Diaspora of Indian origin around the universe unite with the homeland and connect or virtually join the carnival. The rest of the world also gets scopes of virtual participation in celebration and developing cultural connections with it and its locations. Addictions to round-the-clock photoshoots, selfies, and groupies that grow among offline visitors during the Durga carnival increase sales of smartphones, selfie-sticks, and power banks significantly (Ghosal, 2016).

Various mobile phone and Internet service providers endeavour to create a festive mood among their consumers, offering typical local drum beats (dhāk), bell tones, hymns (mantras), and devotional tunes to set as ringtone or caller/ hello tune amid 4G connection. The trend of sharing n-numbers of photos; videos; selfies; groupies in digital space after clicking at the backdrop of tangible and intangible magnificent carnival displays; cultural performances; idols, facilitate the Goddess, carnival, and location to reach the world virtually right away in real time. Billions of carnival content in the digital space turn to be the prime tourism driver as those motivate millions across the globe to visit the city during the carnival. Thus publicity and marketing of Kolkata and its autumn carnival become more authentic, faster, and more cost-effective than ever before. Since COVID-19 pandemic restrictions push attendees and visitors to participate and experience the carnival digitally in 2020–21, the Durga or autumn extravaganza of Kolkata goes digital overtly. It becomes apparent in the quantitative increase of online festival services arranged by the festival-event stakeholders and virtual carnival tours at the behest of tourism professionals, news channels, you-tuber, and media companies. The most exciting happening of last decade's Kolkata carnival is Durga's simultaneous awakening and coexistence in the myth and on the earth and digital space with her entry to the virtual world facilitated by her enthusiastic digital generation followers.

A crucial transformation in the festival ecosystem appears with the emergence of co-evolving Web relationships or value networking (Allee, 2003) among festival-event organizers, festive-centric tourism operators, IT and digital companies, and visitors proactively using ICT. It empowers digitally armed visitors to become the vanguard of changes that influence the festival-event stakeholders to modernize the cultural displays and develop a pragmatic promotion, marketing, and operation process. Thus, blurring the myth of irreplaceable divides between festival-event organizers and visitors, the changing dynamics put both on the same level since both participate in developing a new image of the carnival collectively while interacting through ICT-based media (Ray, 2021). A shift from the twentieth-century offline promotion tradition assisted by printed material to experience-driven Web content like the audio podcast or video clips aided online promotion turns instrumental in shaping and creating authentic carnival images globally. Newly emerged Web-based value networking also makes the flow of order, payment, contract, proposal, invoice, confirmation, receipt, information, knowledge, and other know-how transactions cost-effective and easy (Ray, 2021). An incessant sharing of carnival content at the digital space, since the carnival's inauguration, creates scopes for visitors to choose the location for the offline visits and for devotees to buy offerings, God's grace, and gifts as per their choice, needs, and budget. Above all, online opinions created by the digitally armed visitors continuously keep shaping and reshaping, modifying and altering the carnival experience, imaging, promotion, display patterns, and marketing patterns.

Conclusion

Interaction between technology application and Kolkata's autumn extravaganza becomes crucial to transform the Durga festival from traditional state to orderly organized carnival state, growth era to mobility era and then to sustainable practice, and eventually to digital space. A vital disruption occurs with the spread of digital technology and ICT application in festival-event's periphery when it directs the shifts in the paradigm of experiencing imaging, marketing, promotion, display patterns, and operation processes of the carnival. A change in the imaging or portrayal technique and style, image building and branding of Kolkata's autumn extravaganza become evident overtly with the growing pursuit of sharing and uploading digitally armed visitors' created carnival content, mostly videos and photos, on the walls of ICT-powered media. The digital push plays the role of a great unifier as it has unified cross-generation, cross-community, and cross-class, cross-culture dwellers of the city and Bengalis across the world. The spreading of this phenomenon with increasing mobility, of late, attaches the Bengali Diaspora of Indian origin around the universe to their homeland with great emotion, unites and connects them with their counterparts living in the state, allowing them to virtually join the carnival. The rest of the world also has an opportunity to be connected to the celebration virtually. It thus contributes to the increase of festival-centric tourism in the city in recent times. The ICT application-aided value networking system of imaging carnival, creating and formulating offers, solving service issues when shared by visitors, festive-event organizers, tourism stakeholders, corporate, and IT companies, disrupts the existing festival ecosystem to develop a new ecology that blurs the hierarchical hegemony. Thus, digital and ICT interference leads to change the visitor-organizer role-relationship trajectory from top-down to bottom-up.

References

Allee, V. (2003). *The Future of Knowledge: Increasing Prosperity through Value Networks.* Boston, MA: Butterworth-Heinemann.

Allen, J., O'Toole, W., Harris, R., and McDonnell, I. (2010). *Festival and Special Event Management* (5th ed.). Brisbane: John Wiley and Sons.

Arora, S., and Sharma, A. (2020). Digital marketing for religious event of India for tourism sustainability and promotion. In A. Hassan and A. Sharma (Eds.), *The Emerald*

Handbook of ICT in Tourism and Hospitality. Bingley: Emerald Publishing Limited, pp. 453–465.

Backman, K.F., Backman, S.J.U., Uysal, M., and Sunshine, K.M. (1995). Event tourism and examination of motivations and activities. *Festival Management and Events Tourism*, 3, pp. 15–24.

Banerjee, A. (2019) The Puja economy. *Businessworld*. Retrieved from: www.businessworld. in/article/The-Puja-Economy/20-09-2019-176487/ (accessed 28 October 2021).

Breukel, A., and Go, F.M. (2009). Knowledge-based network participation in destination and event marketing: A hospitality scenario analysis perspective. *Tourism Management*, 30(2), 184–193.

British Council (2019). *Mapping creative economy around Durga Puja.* Retrieved from: www.britishcouncil.in/programmes/arts/Mapping-Creative-Economy-around-DurgaPuja (accessed 25 October 2021).

Buczkowska, K. (2009). Kulturowa turystyka eventova. In K. Buczkowska and A. Mikos von Rohrscheidt (Eds.), *Współczesne formy turystyki kulturowej Tom I. Poznań.* Warszawa: Akademia Wychowania Fizycznego im. Eugenusza Piaseckiego w Poznaniu, pp. 91–118.

Chang, J. (2006). Segmenting tourists to Aboriginal cultural festivals: An example in the Rukai tribal area, Taiwan. *Tourism Management*, 27, pp. 1224–1234.

Crompton, J., and McKay, S. (1994). Measuring the economic impact of festivals and events: some myths, misapplications and ethical dilemmas. *Festival Management and Event Tourism*, 2(1), 33–43.

Crompton, J.L., and McKay, S. (1997). Motives of visitors attending festival events. *Annals of Tourism Research*, 24(2), 425–439.

Cudny, W. (2011). Film festivals in Łódź as a main component of urban cultural tourism. *Bulletin of Geography. Socio-economic Series*, 15, 131–141.

Cudny, W., and Rouba, R. (2011). Theatre and multicultural festivals in Lodz as a free time management factor among the inhabitants and tourists in the post-industrial city. *Acta Geographica Universitatis Comenianae*, 55, 3–22.

Cudny, W., Korec, P., and Rouba, R. (2012). Resident's perception of festivals – the case study of Łódź. *Slovak Sociological Review*, 44, 704–728.

De Bres, K., and Davis, J. (2001). Celebrating group and place identity: A case study of new regional festival. *Tourism Geographies*, 3, 326–337.

De Valck, M. (2007). *Film Festivals: From European Geopolitics to Global Cinephilia.* Amsterdam: Amsterdam University Press.

Djumrianti, D. (2018). The roles of social media in the promotion of traditional cultural tourism events in Indonesia. In A. Hassan and A. Sharma (Eds.), *Tourism Events in Asia*. Abingdon: Routledge, pp. 114–122.

Falassi, A. (1987). *Time Out of Time: Essays on the Festival.* Albuquerque, NM: University of New Mexico.

Ferdinand, N., and Williams, N. L. (2013). International festivals as experience production systems. *Tourism Management*, 34, 202–210.

Formica, S., and Uysal, M. (1996). A market segmentation of festival visitors: Umbria Jazz Festival in Italy. *Festival Management and Event Tourism*, 3(1), 75–82.

Getz, D. (1991). *Festivals, Special Events, and Tourism.* New York: Van Nostrand Rheinhold.

Getz, D. (2000). Festivals and special events: Life cycle and saturation issues. In W. Gartner and D. Lime (Eds.), *Trends in Outdoor Recreation, Leisure and Tourism.* Wallingford: CABI, pp. 175–185.

Getz, D. (2008). Event tourism: Definition, evolution, and research. *Tourism Management*, 29(3), 403–428.

Getz, D. (2010). The nature and scope of festival studies. *International Journal of Event Management Research*, 5, 1–47.

Ghosal, A. (2016). Devi's digital domain. *Hindustan Times*. Retrieved from: www.hindustantimes.com/kolkata/devi-s-digital-domain/story-3YPPeidaa 0YY7PZ2qPrUhL.html (accessed 28 October 2021).

Gotham, K.F. (2005). Theorizing urban spectacles: Festivals, tourism and the transformation of urban space. *City*, 9, 225–246.

Greenwood, D. (1972). Tourism as an agent of change: A Spanish Basque case study. *Ethnology*, 11, 80–91.

Gunn, C. (1979). *Tourism Planning*. New York: Crane Russak.

Handiqui, K. (2011). *Somadevas Yasatilaka: Aspects of Jainism Indian Thought and Culture*. New Delhi: DK Printworld.

Janiskee, R. (1980). South Carolina's harvest festivals: Rural delights for day tripping urbanites. *Journal of Cultural Geography*, 1(Fall/Winter), 96–104.

Kowalczyk, A. (2001). *Geografia turyzmu*. Warsaw: Wydawnictwo Naukowe PWN.

Lee, C., Lee, Y., and Wicks, B.E. (2004). Segmentation of festival motivation by nationality and satisfaction. *Tourism Management*, 25(1), 61–70.

Markova, B., and Boruta, T. (2012). The potential of cultural events in the peripheral rural Jesenicko Region. *AUC Geographica*, 47(2), 45–52.

Markwell, K., and Waitt, G. (2009). Festivals, space and sexuality: Gay Pride in Australia. *Tourism Geographies*, 11, 143–168.

Matheson, C.M. (2005). Festivity and sociability: A study of a Celtic music festival. *Tourism Culture and Communication*, 5, 149–163.

McDermott, R.F. (2001). *Mother of my heart, daughter of my dreams: Kali and Uma in the devotional poetry of Bengal*. Oxford: Oxford University Press.

Mika, M. (2007). Formy turystyki poznawczej. In W. Kurek, R. Faracik, M. Mika, W. Pawlusiński, E. Pitrus, and D. Ptaszycka-Jackowska (Eds.), *Turystyka: Warszawa*. Warsaw: Wydawnictwo Naukowe PWN, pp. 198–231.

Mohr, K., Backman, K.F., Gahan, L.W., and Backman, S.J. (1993). An investigation of festival motivations and event satisfaction by visitor type. *Festival Management and Event Tourism*, 1(3), 89–97.

Mossberg, L., and Getz, D. (2006). Stakeholder influences on the ownership and management of festival brands. *Scandinavian Journal of Hospitality and Tourism*, 6(4), 308–326.

Niyogi, S., and Mukherji, U. (2019). Not just fun, Kolkata's Durga Puja. *Times of India*. Retrieved from: http://timesofindia.indiatimes.com/articleshow/71549101.cms?utm _source=contentofinterestandutm_medium=textandutm_campaign=cppst (accessed 15 October 2021).

Phipps, P., and Slater, L. (2010). *Indigenous Cultural Festivals: Evaluating Impact on Community Health and Wellbeing*. Melbourne: Royal Melbourne Institute of Technology.

Picard, D., and Robinson, M. (2006) (Eds.). *Festivals, Tourism and Social Change: Remaking Worlds*. Clevedon: Channel View.

Prensky, M. (2001a). Digital natives, digital immigrants Part 1. *On the Horizon*, 9(5), 1–6.

Prensky, M. (2001b). Digital natives, digital immigrants Part 2: Do they really think differently? *On the Horizon*, 9(6), 1–6.

Prentice, R., and Andersen, V. (2003). Festival as creative destination. *Annals of Tourism Research*, 30, 7–30.

Quinn, B. (2009). Festivals, events, and tourism. In T. Jamal and M. Robinson (Eds.) *The SAGE Handbook of Tourism Studies*. London: SAGE Publications Ltd., pp. 483–504.

Quinn, B. (2010). Arts festivals, urban tourism and cultural policy. *Journal of Policy Research in Tourism, Leisure and Events*, 2, 264–279.

Ray, S. (2018). Event tourism in Asian context. In A. Hassan and A. Sharma (Eds.) *Tourism Events in Asia: Marketing and Development*. Abingdon: Routledge, pp. 4–19.

Ray, S. (2021). a pragmatic approach of interaction between technology and tourism-hospitality. In A. Hassan and A. Sharma (Eds.), *The Emerald Handbook of ICT in Tourism and Hospitality*. Bingley: Emerald Publishing Limited, pp. 19–29.

Ray, S. (in press). Tracing the journey of event-tourism: o-evolution to synergic merger. In E. Dhariwal, S. Arora, A. Sharma, and A. Hassan (Eds.), *Event Tourism and Sustainable Community Development: Advances, Effects and Implications*. Palm Bay, FL: Apple Academic Press Inc.

Schneider, I.E., and Backman, S.J. (1996). Cross-cultural equivalence of festival motivations: A study in Jordan. *Festival Management and Event Tourism*, 4, 139–144.

Scotinform Ltd. (1991). *Edinburgh festivals study 1990/91: visitor survey and economic impact assessment. Final Report*. Edinburgh: Scottish Tourist Board.

Scott, D. (1996). A comparison of visitors' motivations to attend three urban festivals. *Festival Management and Event Tourism*, 3, 121–128.

Shabnam, S., Ramkissoon, H., and Choudhury, A. (2018). Role of ethnic cultural events to build an authentic destination image: A case of 'Pohela Boishakh' in Bangladesh. In A. Hassan and A. Sharma (Eds.) *Tourism Events in Asia: Marketing and Development*. Abingdon: Routledge, pp. 47–63.

Small, K., and Edwards, D. (2003). Evaluating the socio-cultural impacts of a festival on a host community: A case study of the Australian festival of the book. In T. Griffin and R. Harris (Eds.), *Proceedings of the 9th annual conference of the Asia Pacific tourism association*. Sydney: University of Technology, pp. 580–593.

Snowball, J., and Willis, K. (2006). Building cultural capital: Transforming the South Africa National Arts Festival. *South Africa Journal of Economics*, 74 (1), 20–33.

Statista (2021a). *Number of internet users in India from 2010 to 2020, with estimates until 2040 (in millions)*. Retrieved from: www.statista.com/statistics/255146/number-of-internet-users-in-india/ (accessed 15 October 2021).

Statista (2021b). *Number of social network users in India from 2015 to 2020, with estimates until 2040 (in millions)*. Retrieved from: www.statista.com/statistics/278407/number-of-social-network-users-in-india/ (accessed 15 October 2021).

Turner, V. (1982). Introduction. In V. Turner (Ed.), *Celebration: Studies in Festivity and Ritual*. Washington, DC: Smithsonian Institution Press, pp. 11–29.

Vu, H.Q., Li, G., and Law, R. (2018). Cross country analysis of tourist activities based on venue: Referenced social media data. *Journal of Travel Research*, 59(1), 90–106.

Vu, H.Q., Li, G., Law, R., and Zhang, Y. (2017). Tourist activity analysis by leveraging mobile social media data. *Journal of Travel Research*, 57(7), 883–898.

Xie, P.F. (2003). The Bamboo-beating dance in Hainan, China: authenticity and commodification. *Journal of Sustainable Tourism*, 11(1), 5–16.

5 Kaamatan Goes Virtual

Utilizing Social Media in Promoting Tourism Event

*Sharifah Nurafizah Syed Annuar and
Cynthia Robert Dawayan*

Introduction

Festivals and events have long existed in human society and were presumed as a form of ritual, traditions or collective celebration. They serve as social activities which allow a certain community to gather, celebrate certain causes as well as to exhibit and showcase a wide variety of products and services. Generally, festivals and events are unique due to their interaction with the people, environment and activities involved. Their ability to give individuals the opportunity to experience and fulfil certain needs are often regarded as an important pull factor in tourism industry. With this in mind, most countries draw on festivals and events to boost destination image, attract visitors and create recreational opportunities for potential tourists.

Social Media Usage in Tourism Industry

In January 2021, the global daily social media usage was approximately 4.2 billion or 53.6% of the global population (Chaffey, 2021). It has become one of the most popular digital activities that all levels of society around the world engages in (Tankovska, 2021). This number is expected to increase over the years as more and more people get better access to the Internet as infrastructures become better and mobile devices become more accessible to most people (Minazzi, 2015). Availability of affordable mobile phones and Internet packages has stimulated the growth of social media users where it was recorded that 4.15 billion (98.8%) users access social media through their mobile devices (Chaffey, 2021). Facebook was the most preferred social media platform used (2.74 million), followed by YouTube (2291 million), WhatsApp (2000 million), Facebook messenger (1300 million) and Instagram (1221 million) (Chaffey, 2021). It is through these platforms that users can perform various tasks, receive important updates and other useful information. Facebook is not just a platform to update status and share pictures with friends, it also provides other interactive and transactional features such as FB Marketplace. Additionally, Facebook also serves as an effective tool to promote and create awareness among the community about events and

DOI: 10.4324/9781003271147-8

other happenings around them. The same occurs with other platforms such as YouTube and Instagram, where contents related to a particular event, product or service are being shared in the platforms with the aim to inform, educate and attract society about the event, product or service (Jorge et al., 2020).

Within the tourism industry, many tourism players have started to actively incorporate the use of social media in promoting their tourism products as well as communicate with their past, current and potential tourists. They learned that social media was the most effective platform that helps them reach out to their tourists in a more interactive manner with no geographical boundaries and can be updated regularly. Social media has also become the "go to" place for tourists to find information when they are gathering information before making any vacation decisions. Tourists' vacation decisions are largely be influenced by the information, recommendations and comments from previous tourists available in the various social media sites (Werenowska and Rzepka, 2020). Additionally, social media is also used by tourism players to boost their brand and the destination that they are promoting by providing relevant information to attract the attention of potential tourists. Therefore, it is important for all tourism players to have a good social media implementation strategy, in order to boost their business by getting to know their tourists' needs and wants better and being able to fulfil those needs better than their competitors. They need to come up with a systematic way to manage their social media accounts to ensure that it will benefit them in terms of not only informing and marketing their products to potential tourists but at the same time to build the organization's image as well as the destination brand itself.

The use of social media has become extremely important as the whole world faces the COVID-19 pandemic. Importance of social media has doubled as the most effective way for people around the world to keep in touch as nations implement travel restrictions and people all over the world have to go into lockdowns and stay at home. Almost everything was put on standstill in the effort to stop the spread of the virus. Industries and businesses around the world were affected and the tourism industry is among those that are hit the hardest due to travel restrictions and social distancing policies implemented globally. Destinations, hotels, resorts, attraction parks and almost all tourism-related activities were forced to close. This is the largest challenge that tourism players have to face as they find ways to save their business and at the same time to service their loans and other financial commitment. Yet, tourism players are hopeful that the pandemic will pass and tourists will return. Therefore, it is important for them to continuously market their products and services to the world. They need to continuously communicate and maintain their relationship with their previous and potential tourists. Some establishments have initiated virtual tours around their establishment for their social media followers just to keep in touch with them. Other tourism establishments organize online activities to engage with their followers, while others continuously promote their services by having special packages while ensuring that COVID-19 health protocol is being observed as tourism activities slowly resume. Generally, COVID-19 has not only changed the landscape of the tourism industry but also how tourism marketing and tourists' decision making are made (Toubes et al., 2021).

Social media and tourism events

The tourism industry has been inevitably affected by the Internet and has become the lifeblood of the industry. Information searching of target destinations, travel planning in real time and bargaining packages are among the examples of how tourists use the Internet to travel. In the same manner, service providers have received many benefits of using the Internet too. It enables them to redistribute their services through online intermediaries, employ omni-channels to promote their packages and services and offer better customer experience to the potential users. The rise of social media – for instance Facebook, Instagram, YouTube, Twitter, and TikTok – have given the tourism industry advantages in pulling tourists to their destinations.

Apart from the typical promotions of products and services in the tourism industry, service providers and related agencies recognize the significance of social media in attracting tourists to visit their destinations by highlighting relevant cultural events and festivals. Social media is commonly used to promote festivals and events, particularly among young audiences who are more likely to appreciate the use of multiple social media channels. Social media is presently the best way to engage with audiences as it can generate better results compared to other traditional tools. In addition, its capability to provide real-time information through live sessions and two-way communications at minimal costs make social media more appealing. The use of social media in promoting tourism events are also remarkable. Previously, social media was used to promote tourism events and to encourage tourists to visit a destination. However, with the COVID-19 pandemic going on, event managers optimize the features of social media to broadcast their events. In this sense, social media enables them to gain a wider audience with the hope of encouraging potential tourists to attend the events in the future.

From the consumers' perspective, they are keen to experience services that elicit emotional engagement which explains the behaviour of information search and making decisions with the aid of social media. The tourism industry should exploit social media as the platform that allows participation and interactions with target consumers. In this regard, consumers play the roles of co-marketer to others' travel experiences. They think they are making a major contribution to the decision making of others and this makes them feel good about it. Therefore, noting the vital impact of social media in collaborative marketing, the platform should be construed as one of the significant tools in promoting festivals.

Case: Harvest Festival (*Kaamatan*) in Sabah, Malaysia

In 2019, Malaysia welcomed a total of 26.1 million inbound tourists, with total receipts of RM86.1 billion, with the tourism industry contributing 15.9% to the country's gross domestic product (GDP) (Department of Statistics Malaysia, 2020). The top five countries contributing to the Malaysian tourist arrival included Singapore, Thailand, Indonesia, Brunei, and China. Additionally, 239.1 million domestic tourists were recorded in 2019, contributing a total of RM103.2 billion in domestic expenditure.

In Sabah, 2019 records a total of 1,469,475 international tourists and 2,726,428 domestic, contributing a total of RM 9.01 billion to the state's income, with the main group of tourists being from China and South Korea (Sabah Tourism, 2020). These groups of tourists are keen on participating in cultural and environmentally based activities that offer them the opportunity to learn and gain new experiences. Usually, there is a surge in the number of tourist arrivals to Sabah in the month of May as tourists from all over the world take the chance to visit Sabah to witness for themselves the unique Harvest Festival.

Kaamatan, better known as Harvest Festival, is the highlight of all events in Sabah, Malaysia. It is a cultural event that is typically celebrated by one of the largest ethnic groups in Sabah which is the *Kadazandusun*. Celebrations stretch throughout the month of May leading to the highlight of the celebration which is on 30 and 31 May each year. The celebration revolves around remembering and thanking God (*Kinoingan*) for his sacrificial love for mankind, when he sacrificed his only daughter (*Hominodun*) to save the human race. According to the local beliefs, *Huminodun was* sacrificed during a long season of drought and shortage of paddy. Parts of her body were buried in the fields and subsequently grew as paddy (Topin, 2015). To commemorate the sacrifice, the locals perform rituals such as *Magavau* to make peace with paddy spirits *Bambarayon* during this festival.

Apart from performing rituals, the *Kadazandusun* community will prepare a variety of traditional delicacies and *Tapai* (local rice wine) to be shared among family members and guests. Additionally, traditional songs will be aired to which they will perform cultural dances, play traditional games and sports. Furthermore, there is also the must-have *Unduk Ngadau* (beauty pageant) and *Sugandoi* (singing) competition. During the two-day highlight of the event, held at *Hongkod Koisaan* in *Penampang*, Sabah, visitors can see an array of ethnic food, drinks and handicrafts being put up for sale. Additionally, visitors also can participate in traditional games and making of traditional handicrafts. This vibrant celebration is the time of the year where the *Kadazandusun* community gather to celebrate and bring back to life traditional rituals and activities that are slowly being forgotten by the younger generation. It has also become a selling point to attract tourists to visit Sabah and attend the event. It is one of the most anticipated events by tourists from around the world, resulting in them scheduling their visit to Sabah during this time of the year.

However, due to the COVID-19 pandemic, many events were either postponed or cancelled as the holding of these events was against the COVID-19 health policy and also the large gathering restrictions that were imposed around the world. Among the prominent events postponed were the 2020 Summer Olympics, UEFA Euro 2020 and Copa América 2020 which were been postponed to 2021, while some events were postponed indefinitely (Westmattelmann et al., 2021). In Malaysia, among the events cancelled in 2020 were Shell Malaysia Motorcycle Grand Prix 2020 and EcoWorld's Women Summit 2020. Other than that, the most awaited Visit Malaysia 2020 Campaign was cancelled indefinitely as there was obviously no way for the campaign to proceed due to the closure of non-essential businesses and travel restrictions imposed in Malaysia (Tourism Malaysia, 2020). Events in Sabah were also not spared, when organizers had to cancel the ANZAC Memorial event held annually to commemorate the death of Australian and New Zealand prisoners of war as well as the most awaited Harvest Festival Event.

In 2021, most event organizers had taken the initiative to modify the celebration and holding of events to suit COVID-19 health policies. Events were held without any live audience and can be viewed only via live streaming. The Harvest Festival Celebrations 2021 in Sabah were carried out via live streaming, where only a limited number of committee members and participants were allowed to be on site, as the event is being streamed live via social media platforms such as Facebook and YouTube. Themed as *"Borderless Kaamatan"*, live streaming was done for the *Sugandoi* as well as *Unduk Ngadau* events. Various social media platforms were utilized to promote the event prior to the celebration which kicked off on 1 May 2021. Information was disseminated via the various social media platforms such as Facebook and Instagram in order to inform society about the initiative and for them to get more information and latest updates about the event. These pre-event postings or information shared was able to stir the interest among followers who look forward to the event. This effort was a success as it opened up opportunities for *Sabahans*, particularly those abroad, to follow the live streaming and celebrate the joys of *Kaamatan* from afar.

Our recent online survey conducted among 118 people residing in Sabah found that more than half (53.4%) of the respondents followed the *Unduk Ngadau Kaamatan* (Harvest Festival Beauty Pageant) 2021 related social media accounts. These accounts document the journey of the participants from the beginning of their participation in the pageant. The followers are excited to see updates about the events as they wait in anticipation for the crowning of the *Unduk Ngadau*. This suggests that the respondents use social media to keep themselves updated on what is going on especially in this trying time. This chapter, therefore, aims to discuss and suggest how to increase motivation to attend the festival by using the social media approach.

Some Suggestions

Motivation to attend festivals

Studies of motivation to attend festivals have drawn considerable attention from academia. A literature review was conducted to get an overview on visitors' motivation to attend festivals. The words "festival motivation", "cultural festival motivation" and "online festival motivation" were employed as keywords to find academic articles from SCOPUS database. Several hundred journal papers were found; however, only related articles from the year 2020 to 2021 were reviewed for the purpose of this chapter. Thus, only eight papers are included in the summary (see Table 5.1).

Our findings found that studies on motivations to attend festivals mostly involve events related to entertainment (e.g., concert), sports, food and beverages (e.g., wine), cultural and online shopping. This indicates that there is a gap in the literature on the motivation to attend online cultural festivals that needs to be filled. In addition, the use of social media to encourage online festival visits is also lacking in the recent literature. Furthermore, the recent studies are mostly performed in qualitative manners, using thematic analysis to understand the motivations to attend festivals.

Contemporary studies on motivations to attend festivals (Table 5.1) discuss factors such as personal motivation, social factors, theme-related event motivations, online community participations, escapism, visitor–environment fit including facilities, environmental functions, and activity knowledge, cultural attractions and utilitarian-hedonic motivations.

Motivation is defined as the drives to prompt a behavioural change to satisfy a need (Westbrook and Black, 1985). The underlying reason for a consumer to purchase and consume a product or service is not only to obtain the functional benefits but also to fulfil their personal and social motives. Personal motive is often associated with responsibility, physical activity, sensory stimulation and gratification (Mehta et al., 2014) while social motive is the experience a consumer gains when interacting with other people during the purchases or consumption (Kumar and Sadarangani, 2021). Fundamentally, scholars have categorized purchase and consumption behaviour based on either utilitarian or hedonic motivation (Kumar and Sadarangani, 2021). These motivations which were traditionally investigated among marketing scholars to determine consumer behaviour were then studied in the area of tourism and hospitality. According to Holbrook and Hirschman (1982), consumption experience should be understood from an experiential perspective. The feelings manifested in the experience together with rational evaluation has stirred researchers to apply the concepts in tourism and hospitality studies (Titz, 2008).

Utilitarian versus hedonic motivations

Utilitarian motivation creates favourable or non-favourable attitudes about certain tasks while hedonic motivation comprises of feelings of happiness, charm or spontaneity (Babin et al., 1994). Consumers who are motivated by utilitarian values are objective, rational and logical and often have predetermined expectations on what they should attain in their purchases or consumption. In relation to social media use, this type of motivation leads to consumers browsing for information online so that they can have some thoughts before making any purchases or consumptions (Kim and Kim, 2004). Additionally, Ramlugun and Jugurnauth (2014) discovered that utilitarian consumers are motivated by convenience and cost-saving when they browse information on social media. Consequently, if

Table 5.1 Earliest attempts to understand motivation to attend festivals

Authors	Theoretical Bases	Antecedents	Key Findings
Zhang et al. (2021)	Social Exchange Theory	General Festival Motivation and Theme-Related Motivation	Visitors' general festival motivations had a favourable impact on their place attachment and satisfaction. Visitors' theme-related motivations had a positive effect on their place identity and satisfaction.
Kabiraj et al. (2021)			While primary and secondary motivations have a positive influence on the perception of food and wine, generic features, and fun; only primary motivation is found to influence satisfaction positive. However, secondary motivation has negative relationship towards satisfaction. Additionally, perception of wine and food and fun has a favourable influence on satisfaction, while the perception of generic features is found to be insignificant.
Muhs et al. (2020)		Social and Cultural Factors	Social factors, such as friendships in vital in encouraging return visits.
Chen and Li (2020)	Stimulus-Response Theory	Perceived Temptation of Price Promotion, Perceived Categories Richness of Promotion and Perceived Fun of Promotion Activities and atmosphere promotion strategies as Perceived Contagiousness of Mass Participation.	Perceived Temptation of Price Promotion, Perceived Categories Richness of Promotion, Perceived Fun of Promotion Activities and Perceived Contagiousness of Mass Participation significantly and has positive effect on consumer Participation Intention; Perceived Contagiousness of Mass Participation moderates the relationship between Perceived Temptation of Price Promotion and Participation Intention.
Zou et al. (2020)			Cultural contact is significant towards festival attendees' future behavioural intention, while visitor–environment fit, including facilities, environmental functions, and activity knowledge, affect attendees' experience of cultural contact.
Choo and Park (2020)			Festival/escape and family togetherness motivation dimensions have favourable impact towards involvement for both local and non-local visitors. While local food motivation impact positively involvement only for non-local visitors, socialization motivation determines local visitors' involvement.
Ďađo et al. (2020)			Both satisfaction and visitors' subjective well-being are found to be direct antecedents to visitors' behavioural intentions and mediators of the impact of perceived value on visitors' behavioural intentions.
Shang et al. (2020)			Behavioural findings showed that participants had a purchase intention bias towards utilitarian products compared to hedonic products.

they encounter exciting and pleasant experience during information search, their hedonic motivations will be activated (Arum and Sung, 2018). When consumers assume an event as practical, and develop liking, this will encourage them to go deeper by analysing brands tied up with the event which results in the formation of utilitarian evaluations. Likewise, if a consumer feels that an event is favourable, this will enable them to create positive feeling towards brand/s that is/are associated with the event and indirectly will lead them to generate liking towards the brand (Sreejesh et al., 2021).

In the case of the *Kaamatan*, to encourage motivation to attend the event, the organizers are advised to look into utilitarian and hedonic values. These types of values would encourage potential visitors to browse information online and facilitate them in making travel decisions. While utilitarianism is associated with functional needs, hedonism is linked with emotional needs. In meeting the functional needs, theory of cognitive evaluation explains that individual extrinsic acts are fulfilled for external benefits and this behaviour can be triggered by offering functional rewards like usefulness of certain products or services. For that reason, *Kaamatan* organizers must identify the functional benefits that they can offer to elicit this motivation. As previously discussed in the Technology Acceptance Model (TAM), Davis (1989) highlighted perceived usefulness and perceived ease of use are the ultimate factors in accepting new technology. With this in mind, the organizers probably need to look into the social media functions and attributes and ensure that the platforms used by the organizers are user-friendly and useful in showcasing the virtual event. If this is observed and met, the utilitarian motivation is satisfied and thus will induce liking towards the event and subsequently encourage the viewers to attend the event in the future.

Additionally, the organizers can also consider working with local and international brands through smart partnership to develop liking towards the event. This is a win-win situation not only for the organizer but also for the brands. The brands can support the organizer by providing prizes for the games or contests during the festival. *Kaamatan* is a good platform for brands to enhance their brand image as the event is culturally unique and diversified. It is not limited to showcasing certain ethnicity rituals but also filled with ethnic cuisines presentations and entertainment elements like a singing competition. This event has also many followers from all ages, and has wider acceptance be it from the locals or foreign tourists due to its cultural distinctiveness. Therefore, it is a good opportunity for the brands to be part of the festival.

Browsing and participation behaviour

Browsing behaviour can be observed from a person's social media information search and consumption activities in obtaining information about their area of interest. This includes activities such as reading articles, infographics, viewing pictures and watching videos (Bateman et al., 2012). Due to unlimited information availability, individuals must screen for the usefulness and relevance of the information. This screening is done based on human perception and personal knowledge (Rowley, 2002). Participation behaviour on the other hand is the behaviour of consumers as they interact with an online community through learning, socializing and exchange of information and opinion (Brodie et al., 2013). Hu, Zhang and Luo (2016) suggested two types of participation behaviours on social media platforms, which are content contributor and community participation. Content contributors refer to community members who actively contribute information through videos, articles and pictures, while community participation refers to social relationships among community members. According to Madupu and Cooley (2010), active participation among community members is essential in order to maintain excitement and sustainability of an online community. Moreover, it also signifies consumer commitment to a particular brand (Dessart et al., 2015).

Studies were also done to examine the relationship between followers' motivations and browsing behaviour, where it was revealed that people with utilitarian motivation are less interested in new websites and are unlikely to search online for the latest information (Cotte et al., 2006). They will only engage in interactive activities if they perceive their participation as beneficial and helpful for them in their information search (Dholakia et al., 2004). A study done based on a travel agency's WeChat official account suggests that hedonic motivation does not indicate a significant relationship with participation behaviour, rather it was the utilitarian motivation that has a positive relationship with participation behaviour (Liang and Yang, 2018). It was found that users prefer to utilize the official account as they perceive that interacting via the account offers them utilitarian benefits as well as provides them with the relevant information needed. On the other hand, research based on social media platforms for leisure-driven events revealed that both utilitarian and hedonic motivation have important effects on browsing and participation behaviours. These behaviours also influence users' intention to share information about the event with their friends as well and to increase their intention to attend an event.

Nonetheless, in promoting cultural-driven events, both hedonic and utilitarian motivations must be given equal attention, where organizers should identify the emotional needs among the followers such as pleasure, joy, fun and curiosity. This can be done by offering a range of games, contests and prizes for the viewers to encourage them to participate in the event. High involvement in the virtual event will generate emotional benefits to the viewers like pleasure and fun which leads to develop liking towards the event. Games and quizzes related to cultures that are unique to the viewers will absolutely create curiosity. They will browse information about the event and participate actively which will in turn encourage them to attend virtually. Functions of the social media platforms encourage users to share the official page to their circle of friends, hence helping to boost the number of followers and increase the visibility of the page. In addition, the ability to send updates or encourage the followers to participate in certain postings through notification buttons will certainly ensure the followers keep track with the updates from the organizers to fulfil their emotional needs.

Conclusion

During the pandemic, festival organizers have no options than to conduct their events virtually through social media platforms. To encourage individuals to attend the event virtually, organizers have to understand the motivations that inspire their potential visitors. From the point of view of the utilitarian and hedonic motivation, efforts concentrated to fulfil the functional and emotional needs must be given careful attention to promote the *Kaamatan*. This is inevitable for the success of the event as organizers will not only attract the public to attend the event virtually but also to motivate them to attend the real event in the future when the borders are open and travel restrictions are eased. This chapter maintains that popularizing this event through meeting the tourist's needs will increase the motivation to attend the event virtually and better promote not only the festival itself but also the destination.

References

Arum, E.S., and Sung, M. (2018). The effect of social media attributes on purchase intention through motivation dimensions and social media product browsing. *Journal of Marketing Thought*, 5(2), 12–23.

Babin, B.J., Darden, W.R., and Griffin, M. (1994). Work and/or fun: Measuring hedonic and utilitarian shopping value. *Journal of Consumer Research*, 20(4), 644–656.

Bateman, S., Teevan, J., and White, R.W. (2012). The search dashboard: How reflection and comparison impact search behavior. In *Proceedings of the SIGCHI Conference on Human Factors in Computing Systems*, pp. 1785–1794.

Brodie, R.J., Ilic, A., Juric, B., and Hollebeek, L. (2013). Consumer engagement in a virtual brand community: An exploratory analysis. *Journal of Business Research*, 66(1), 105–114.

Chaffey, D. (2021). *Global Social Media Research Summary 2021*. Retrieved from: www.smartinsights.com/social-media-marketing/social-media-strategy/new-global-social-media-research/ (accessed 13 August 2021).

Chen, C., and Li, X. (2020). The effect of online shopping festival promotion strategies on consumer participation intention. *Industrial Management and Data Systems*, 120(12), 2375–2395.

Choo, H., and Park, D.B. (2020). Comparison between local and non-local visitors for local food festivals. *Asia Pacific Journal of Tourism Research*, 25(6), 692–705.

Cotte, J., Chowdhury, T.G., Ratneshwar, S., and Ricci, L.M. (2006). Pleasure or utility? Time planning style and web usage behaviors. *Journal of Interactive Marketing*, 20(1), 45–57.

Ďaďo, J., Maráková, V., Táborecká-Petrovičová, J., and Rajić, T. (2020). Modelling the determinants of festival visitors' behavioural intentions. *E&M Economics and Management*, 23(2), 173–190.

Davis, F.D. (1989). Perceived usefulness, perceived ease of use, and user acceptance of information technology. *MIS Quarterly*, 13(3), 319–340.

Department of Statistics Malaysia (2020). *Home*. Retrieved from: www.dosm.gov.my (accessed 10 August 2021).

Dessart, L., Veloutsou, C., and Morgan-Thomas, A. (2015). Consumer engagement in online brand communities: a social media perspective. *Journal of Product and Brand Management*, 24(1), 28–42.

Dholakia, U.M., Bagozzi, R.P., and Pearo, L.K. (2004). A social influence model of consumer participation in network-and small-group-based virtual communities. *International Journal of Research in Marketing*, 21(3), 241–263.

Holbrook, M.B., and Hirschman, E.C. (1982). The experiential aspects of consumption: Consumer fantasies, feelings, and fun. *Journal of Consumer Research*, 9(2), 132–140.

Hu, M., Zhang, M., and Luo, N. (2016). Understanding participation on video sharing communities: The role of self-construal and community interactivity. *Computers in Human Behavior*, 62, 105–115.

Jorge, F., Teixeira, M.S., Fonseca, C., Correia, R.J., and Gonçalves, R. (2020). Social media usage among wine tourism DMOs. In A. Rocha, J.L. Reis, M.K. Peter and Z. Bogdanović (Eds.), *Marketing and Smart Technologies*. Singapore: Springer, pp. 78–87.

Kabiraj, S., Upadhya, A., and Vij, A. (2021). Exploring the factors affecting the behavioral intention of visitors in wine festival: The case of China Dalian International Wine and Dine Festival. *Business Perspectives and Research*, 9(3), 352–369.

Kim, W.G., and Kim, D.J. (2004). Factors affecting online hotel reservation intention between online and non-online customers. *International Journal of Hospitality Management*, 23(4), 381–39.

Kumar, S., and Sadarangani, P. (2021). An empirical study on shopping motivation among generation Y Indian. *Global Business Review*, 22(2), 500–516.

Liang, X., and Yang, Y. (2018). An experimental study of Chinese tourists using a company-hosted WeChat official account. *Electronic Commerce Research and Applications*, 27, 83–89.

Madupu, V., and Cooley, D.O. (2010). Antecedents and consequences of online brand community participation: A conceptual framework. *Journal of Internet Commerce*, 9(2), 127–147.

Mehta, R., Sharma, N.K., and Swami, S. (2014). A typology of Indian hypermarket shoppers based on shopping motivation. *International Journal of Retail and Distribution Management*, 42(1), 40–55.

Minazzi, R. (2015). *Social Media Marketing in Tourism and Hospitality*. Cham: Springer International Publishing.

Muhs, C., Osinaike, A., and Thomas, L. (2020). Rave and hardstyle festival attendance motivations: a case study of Defqon. 1 weekend festival. *International Journal of Event and Festival Management*, 11(2), 161–180.

Ramlugun, V.G., and Jugurnauth, L. (2014). The scope of social media browsing and online shopping for Mauritian e-retailers: a study based on utilitarian and hedonic values. *Review of Integrative Business and Economics Research*, 3(2), 219–241.

Rowley, J. (2002). "Window" shopping and browsing opportunities in cyberspace. *Journal of Consumer Behaviour: An International Research Review*, 1(4), 369–378.

Sabah Tourism (2020). *Statistics*. Retrieved from: h www.sabahtourism.com/statistics/ (accessed 19 August 2021).

Shang, Q., Jin, J., and Qiu, J. (2020). Utilitarian or hedonic: Event-related potential evidence of purchase intention bias during online shopping festivals. *Neuroscience Letters*, 715, 134665.

Sreejesh, S., Sarkar, J.G., and Sarkar, A. (2021). Consumers' responses to tie-in brand purchase intention in event sponsorships. *Event Management*, 25(5), 565–580.

Tankovska, H. (2021). *Social Media – Statistics and Facts*. Retrieved from: www.statista.com/ topics/ 1164 /social-networks/ (accessed 16 August 2021).

Titz, K. (2008). Experiential consumption: Affect-emotions-hedonism. In H. Oh (Ed.), *Handbook of Hospitality Marketing Management*. London: Routledge, pp. 324–352.

Topin, B. (2015). *The Kaamatan Festival*. Retrieved from: www.kdca.org.my (accessed 15 August 2021).

Toubes, D.R., Araújo Vila, N., and Fraiz Brea, J.A. (2021). Changes in consumption patterns and tourist promotion after the COVID-19 pandemic. *Journal of Theoretical and Applied Electronic Commerce Research*, 16(5), 1332–1352.

Tourism Malaysia (2020). *Cancellation of Visit Malaysia 2020 (VM2020) Campaign and Guest Stay as Tourist Accommodation Premises Throughout the Movement Control Order (MCO)*. Retrieved from: www.tourism.gov.my/media/ (accessed 7 August 2021).

Werenowska, A., and Rzepka, M. (2020). The role of social media in generation Y travel decision-making process (case study in Poland). *Information*, 11(8), 396.

Westbrook, R.A., and Black, W.C. (1985). A motivation-based shopper typology. *Journal of Retailing*, 61(1), 78–103.

Westmattelmann, D., Grotenhermen, J.G., Sprenger, M., and Schewe, G. (2021). The show must go on-virtualisation of sport events during the COVID-19 pandemic. *European Journal of Information Systems*, 30(2), 119–136.

Zhang, Y., Park, K.S., and Song, H. (2021). Tourists' Motivation, Place Attachment, Satisfaction and Support Behavior for Festivals in the Migrant Region of China. *Sustainability*, 13(9), 5210.

Zou, Y.G., Meng, F., Li, N., and Pu, E. (2020). Ethnic minority cultural festival experience: Visitor–environment fit, cultural contact, and behavioral intention. *Tourism Economics*, 27(6), 1237–1255.

6 Impact of Online and Social Media Platforms in Organizing the Events

A Case Study on Coke Fest and Pakistan Super League

Syed Arslan Haider, Anum Rehman and Shehnaz Tehseen

Introduction

The aim of this study is to investigate the effect of how social media platforms, e-ticketing systems are used across Pakistan, specifically in the scenario of Coke Fest and Pakistan Super League (PSL). This study will focus on how both events used online ticketing systems during the COVID-19 Pandemic. It will measure the success of the system and the event. Pakistan being a fairly new entry in the e-ticketing business needs to learn from competitors across the world who only use online ticketing systems as modes of selling. This chapter includes the importance and relevance of information technology and communication over the major social websites. These sites include Facebook as a major source of communication for Pakistan's context. This study has been supported by articles, journals, newspapers and government sources to identify the effect of IT and online communications on the events altogether. The use of social media is widespread. Coke Fest and PSL used this to their advantage and advertised their events rigorously over Facebook and Instagram as main sources. Lastly, this chapter aims to find optimal solutions for businesses looking to develop their businesses online. The list of abbreviations for terms used in this chapter is presented as Table 6.1 below:

Information Technology and Information and Communication Technology

What is information technology?

With the advent of the millennium, communications methods drastically changed and evolved. Before the Internet came into existence, traditional methods of communication were used and preferred. However, considering the cost attached to these methods, most companies and organizations started looking for cheaper methods of communication. That's where information technology (IT) came in (Lin et al., 2020). From personal computers to mobile phones and laptops, IT

DOI: 10.4324/9781003271147-9

Table 6.1 List of abbreviations

IT	Information technology
FMCGs	Fast-moving consumer goods
SM	Social media
SMEs	Small and mid-size enterprises
PSL	Pakistan Super League
UAE	United Arab Emirates
HBL	Habib Bank Limited
CF	Coke Fest
SOPs	Standard operating procedures
ICT	Information Communication Technology
VoIP	Voice over IP
GDP	Gross domestic product
NRI	Networked Readiness Index
SNM	Social Networks Marketing
ICCI	Islamabad Chambers of Commerce and Industry
PCB	Pakistan Cricket Board

has revolutionized the way we communicate and manage business operations. Nowadays, information technology management happens to be one of the most important tools used to spread brand messages and engage consumers (Cai and McKenna, 2021). It has helped in shaping the new modes of communication and purchase. Information technology may be advancing but its implications within developing countries are somewhat challenging.

Information and communication technology

In this time and age, every business needs to have an online presence. This does not only imply having a website or social media pages, but it also means having customer engagement across their website and social media pages. Information and communications technology mean just that. According to Buhalis and O'Connor (2005), information communication technology (ICT) has a four-fold meaning,

- Information – (or data) in paper or electronic format
- Communication – in person or electronically (electronic communications), in writing or voice, telecommunications and broadcasting
- Information technology (IT) – including software, hardware and electronics
- Communications technology – including protocols, software and hardware

This definition clarifies that ICT deals with information, its communication over a website or an application with an audience. ICT refers to technologies that provide access to information through telecommunications. Furthermore, people can communicate in real time with others in different countries using technologies such as instant messaging, Voice over Internet Protocol (VoIP) and video-conferencing. Social networking websites like Facebook allow users from all over the world to remain in contact and communicate on a regular basis. Modern information and communication technologies have created a "global village" (Zembylas and Vrasidas, 2005), in which people can communicate with others across the world as if they were living next door. For this reason, ICT is often studied in the context of how modern communication technologies affect society.

The term "global village" means the vast Internet and the applications/software that connect people across the world (Manjarrés et al., 2021). If a company has an online presence, it can connect with anyone anywhere. According to Müller and Antoni (2020), ICT can be defined as a technological means of collecting (inputting/gathering), collating (processing/analysing) and conveying (outputting/transferring) information via technology. Moreover, Ahmed and Le (2021) define ICT as a diverse set of technological tools and resources used to transmit, store, create, share or exchange information. These technological tools and resources include computers, the Internet (websites, blogs and emails), live broadcasting technologies (radio, television and webcasting), recorded broadcasting technologies (podcasting, audio and video players and storage devices) and telephony (fixed or mobile, satellite, Visio/video-conferencing, etc.). All these definitions clearly indicate that ICT is an important tool to use online to communicate with audiences across the world.

Social Media

Role of social media

Social media (SM) is the backbone of most communications in Pakistan (Khan et al., 2021a). Over time, social media has developed itself not only as a source of sharing but also as a source of promoting businesses, selling products and engaging with customers from all around the world (Cartwright et al., 2021). SM is the front line for businesses that have a hybrid system of operations. Social media not only informs about the product or business but also helps the businesses to put forward their core policies and engage in corporate responsibility with lowered costs. Facebook, Instagram and Twitter are social media hubs that aid in promoting businesses and products (Bragg et al., 2020). From small and mid-size enterprises (SMEs) to international conglomerates, social media has the power to make or break their businesses. The reason being that social media is directly related to the consumers who are providing real-time feedback that can either fail a product/service or make it viral. The social media platform usage in Pakistan indicates that most people are using social media at an increasing rate (Qalati et al., 2021).

IT, ICT, SM and Events

How do IT, ICT and SM help in making an event successful?

Changing times require changes within organizations and their structures. In developing countries, the Internet has had a similar effect. It has enabled people to connect and empowered companies to compete globally. In China and India, the Internet has enabled international companies to step into customer-rich markets and expand their reach (Hong and Harwit, 2020). Internet and globalization have enabled companies to share knowledge and use it to make profits. Information technology has enabled developing countries to invest more in medicine, education and public safety. Many government applications and websites help citizens on a daily basis. Taking Pakistan into consideration, IT has been a major help in providing jobs. The IT boom has aided many software houses and international organizations to flourish and expand their businesses. In Pakistan, almost all businesses have an online presence (Nizam et al., 2020). Having a website or an e-commerce-based application helps these businesses to reach out to more markets and expand their customer base. IT has also enabled these companies to capture international markets. From the beauty industry to apparel to sporting goods and fast-moving consumer goods (FMCGs), IT has helped national and international organizations develop a firm online presence and increase sustainability.

Social media being the new entry in the marketing mix helps in making communications smooth and easy. For every business that has its own online presence, it is mandatory that the website be linked with social media platforms to share information (Briciu and Briciu, 2021). Many businesses use social media to promote their events, activities and products by linking it with their websites. This link makes a bridge between social media platforms and information technology which in turn favours organizations. Most companies utilize their social media presence to promote an event or service. They use this platform to reach out to millions of users and use targeted ads for their campaigns.

The Context of Pakistan

Technological advancements in Pakistan

Digital advancements in Pakistan have increased in the last few years. Many companies have decided to run their operations online in order to reach national and international customers. According to Digital Pakistan, Pakistan introduced its first "Digital Pakistan policy" back in 2018 (Iqbal et al., 2020). The primary aim of this policy was to strengthen the IT industry by building a digital ecosystem. Taking a step forwards, the government of Pakistan launched a "Digital Pakistan Vision" in December 2019 with an aim of enhancing connectivity, improving digital infrastructure, increasing investment in digital skills, promoting innovation and tech entrepreneurship. Giving due importance to IT and its development in Pakistan is of vital importance. According to Nizam et al. (2020), the current government has placed its bets on flourishing IT in Pakistan as it is the future. A comprehensive program has been worked out and launched for building a knowledge-based economy by integrating science and technology with economic development programs. The government has raised the financial commitment with the IT sector and launched a vast number of projects that fall under other ministries but that involves the effective use of science and technology for economic growth.

IT and ICT advancements in Pakistan

According to Invest Pakistan (2020), the actual size of the 2016 digital economy was US$11.5 trillion globally, which was 15.5% of the global gross domestic product (GDP). The base digital assets comprise one-third, or US$3.8 trillion, while digital spillover effects account for the remaining two-thirds, US$7.5 trillion. Pakistan, which has about 60% of its 200 million population in the 15 to 29 age group, represents an enormous human and knowledge capital. Pakistan has more than 2000 IT companies and call centres and the number is growing every year. Also, more than 300,000 English-speaking IT professionals with expertise in current and emerging IT products and technologies, 13 software technology parks, more than 20,000 IT graduates and engineers are being produced each year coupled with a rising start-up culture. In accordance with Pakistan Vision 2025 and the digital policy of Pakistan 2018, the ICT industry size is targeted to reach US$20 billion by 2025. According to the Global Innovation Index (2019), as shown in Figure 6.1, Pakistan performs the best in knowledge and technology outputs and its weakest performance is in infrastructure (Tirmizi et al., 2020).

Pakistan is ranked at 112 out of 143 countries, in Networked Readiness Index (NRI) (Iqbal et al., 2020). There are 10 key indicators to measure NRI as presented in Figure 6.2. Unfortunately, Pakistan is far behind in this and the country ranks better in affordability and Business and Innovation environment. Pakistan is also lagging behind in Skills and Infrastructure which are vital for ICT success.

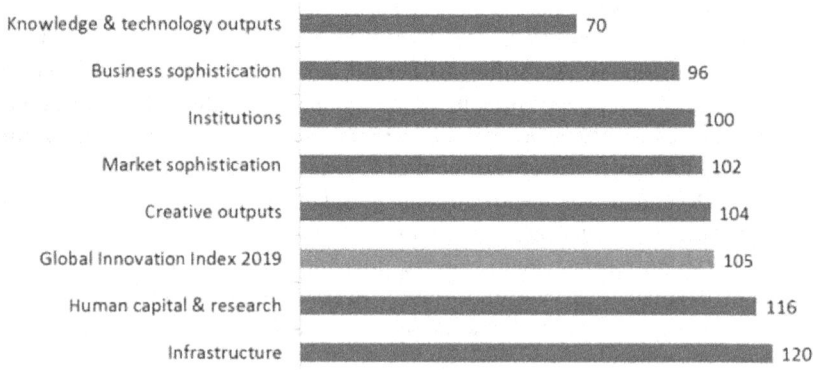

Figure 6.1 Ranking
Source: Cornell University, INSEAD and WIPO, 2019

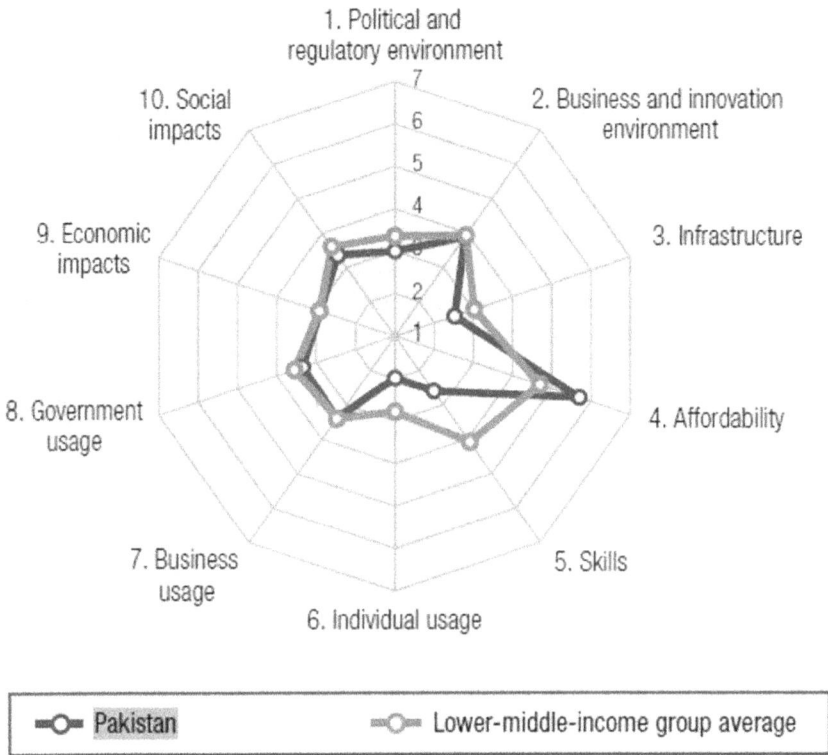

Figure 6.2 10 key indicators of Networked Readiness Index
Source: Zeb, 2015

Pakistan can become one of the top nations in the ICT industry; it will be hard but not unattainable. The future is in ICT and the government must understand this notion and devise the country's digital strategy. Capitalists should also need to think about technology and its future, they should invest in technology to secure the future of their coming generations.

According to World Trade Organization (WTO) (2020), the current coronavirus pandemic and the resulting lockdowns and social distancing measures have had a significant impact on shopping behaviours. As a result, some Business to Consumer (B2C), as well as Business to Business (B2B) firms have seen a significant uptick in sales and e-commerce activity. According to Organisation for Economic Co-operation and Development (OECD) (2020), the evolution of the IT industry combined with the COVID-19 crisis has accelerated an expansion of e-commerce towards new firms, customers and types of products. It has provided customers with access to a significant variety of products from the convenience and safety of their homes and has enabled firms to continue operation in spite of contact restrictions and other confinement measures.

Xinhua (2021) states that as COVID-19 is on the rise, online merchants are increasing too as many businesses are shifting their focus to selling online. Pakistan is among the countries where the e-commerce industry has been steadily on the rise and the number of registered e-commerce merchants has grown significantly over the past few years. After the outbreak of COVID-19, the industry touched new heights of success as people started adopting contactless buying and selling modes. According to the latest data released by the State Bank of Pakistan 2020, the country's e-commerce market has seen a year-on-year growth of 78.9% in volume and 33.3% in value in the fiscal year 2020 (Asghar et al., 2020). Therefore, e-business supports Pakistan to face COVID-19 infectious diseases pressure, because consumers are moving towards electronic commerce (Khan et al., 2021b). By enabling people to conduct many regular activities remotely, including working, learning, shopping and receiving medical services, technology has allowed the continuation of some semblance of a normal lifestyle in this new environment. Within this context, the IT sector has provided the tools and resources required to support these remote activities.

Role of SM in Pakistan

According to Merriam-Webster, social media is defined as, "Forms of electronic communication (such as websites for social networking and microblogging) through which users create online communities to share information, ideas, personal messages and other content (such as videos)" (Merriam-Webster, 2021). Traditionally, consumers used the Internet to simply expend content: they read it, they watched it and they used it to buy products and services. Increasingly, however, consumers are utilizing platforms, such as content sharing sites, blogs, social networking, to modify, share and discuss Internet content. This represents the social media phenomenon, which can now significantly impact a firm's reputation, sales and even survival. Social media are fundamentally changing the way we communicate, collaborate, consume and create. They represent one of the most transformative impacts of information technology on business, both within and outside firm boundaries.

Social media helps in making social media networks which in turn helps to capture consumers from all over the globe. Pentina, Koh and Le (2012) illustrated that social networks marketing (SNM) is strongly influenced by social influences from experts, competitors and customers. These social influences affect intention to adopt this new technology both directly and by affecting the perceptions of the technology's usefulness. For SMEs already using SNM, social influence is the only strong determinant of the intention to continue employing this marketing technology, with the amount of experience with SNM strengthening this relationship.

Since the world communicates through the eyes of social media, tourism can also be promoted by it (Ida and Saud, 2020). Aftab and Khan (2019) explain that various social media platforms are full of pictures and stories of people describing their experiences about the places they have visited. Second, the governmental organizations also launch various social media office profiles to showcase the destinations. Social media is an attractive, informative, useful and approachable way to get information. In the last few years, there has been an increase observed in the smartphone, smart tablet and wireless broadband market in Pakistan. It is because of the popularity of social media, its access and usage in most of the country. It is a positive prospect for the country; however; there are many issues arising with the usage of social networking sites (Memon et al., 2015). Figure 6.3 indicates how social media users in Pakistan divide their time over social media applications. These applications have a huge impact on purchasing, communication and selling of products and services online in Pakistan.

In Pakistan, social media and information technology industries are rising higher than any other industry. Since most businesses have international head offices, they are remotely present only by websites and social media. To promote events, many companies utilize Facebook and Instagram to promote their event. These posts usually have links leading back to the website and promotional material. Pakistan is no stranger to online purchasing. Many companies use online ticketing as a mode of selling products or event tickets. This has become a popular and reliable method to sell products and services. Since the COVID-19 pandemic, many companies have completely gone online to reduce costs. Most companies are hosting online events as a way to retain their customer base. The only disadvantage of designing and implementing an online program is the operational systems may be flawed and might not be optimally implemented.

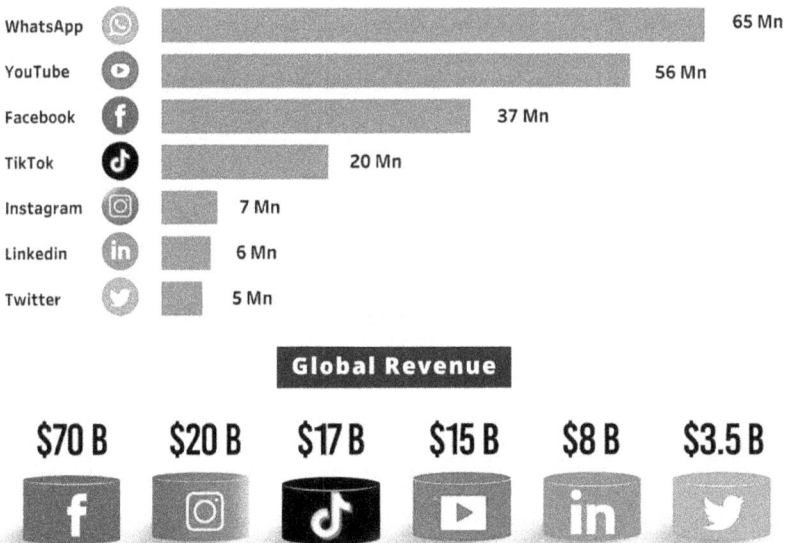

Figure 6.3 Social media users in Pakistan
Source: Islamabad Policy Research Institute, 2020

Coke Fest and Pakistan Super League: Event examples of IT, ICT and SM usages

Conducting an online system of ticketing by utilizing social media platforms and their website is the forte of Coke Fest (CF) and Pakistan Super League (PSL) that was organized and conducted in Pakistan. PSL is a Twenty20 cricket league that was founded in 2015 and made its debut in 2016 in United Arab Emirates (UAE). It consists of six teams representing six cities of Pakistan. The league is managed and monitored under the Pakistan Cricket Board in Lahore. The owner of Arif Habib Group has estimated that PSL's market value was US$300 million and that this event has contributed a lot to the economy and tourism in Pakistan (Aftab and Naveed, 2020). Being such a big event, PSL has a lot of sponsors. Habib Bank Limited (HBL) Pakistan has been sponsoring the event since 2015. HBL also launched a campaign named HBL PSL Hamaray Heroes (HBL PSL Our Heroes) in 2020 and also handed out awards under the same campaign. This campaign was run online and received a huge response. The tickets were sold online through HBL branches. However, due to the spread of COVID-19, PSL had to be postponed and players were sent back to their respective countries (Sattar et al., 2020). All tickets that were sold online and through HBL were returned. By using this hybrid method of selling tickets, PSL has expanded its reach to its consumers.

Coke Fest is a branch of Coca-Cola that was organized to create the first-ever fully digital music event in Pakistan. Coca-Cola Corporation believes in utilizing the latest technologies to deliver food and music to its consumers. Coke Fest was launched in 2017 in Pakistan. It incorporated deals with various bands and singers who would come to perform for the audiences. Alongside music, the organizers had food stalls where different startups would display their products for sale. Coke Fest has always been a successful three-day event across Pakistan. The tickets for the event used to sell through various outlets including institutions. The campaign was carried out using digital media and billboards across Pakistan. Coke Fest was the biggest hyped program in Pakistan. Last year, however, due to COVID-19, Coke Fest could not gather larger crowds and follow standard operating procedures (SOPs). It was then decided to take an unconventional path to entertain people. Coke collaborated with Tapmad, a local video streaming platform; Patari, a leading music application; and FoodPanda, a leading food delivery service provider, to execute a three-day fully digital live concert festival. The concept was to watch and listen to music and bands perform from home and order food which would be delivered to the listeners' doorstep. The idea was to make users download the Patari app and use the unique code to view the live performances of the musicians.

Any company that is being run online needs to have an organized operations system without any bottleneck. Especially in this COVID-19 era, having an online presence is mandatory for any company to survive. This is exactly what Coke Fest and PSL have done. During COVID-19, many cities of Pakistan were under complete lockdown. Only a few citizens were allowed to move around for basic necessities. PSL had collaborated with HBL which had specifically opened counters at its branches to sell HBL tickets. This was done by following proper SOPs implemented by the Government of Pakistan. During the PSL 2020, COVID-19 cases had increased leading the Government to shut down all events in Pakistan and the PSL players leaving for their home countries. Since the matches were cancelled in the remaining cities, PSL had to refund the tickets. For this, PSL had contracted with courier companies for ticket refunds where people who had bought in the tickets could visit the branch and get their refund. This was done by following proper SOPs.

Coke Fest, however, was a completely different scenario. Coke Fest was conducted in collaboration with Tapmad and Patari. Both companies are run via online platforms and have a solid operational system. Patari being a music platform and Tapmad being a live streaming platform collaborated with Coke to produce the first-ever digitally organized event in Pakistan. The system included consumers downloading the Patari application, paying a small fee for registration and then getting a unique code which they could use at the Tapmad website to view all the events and performances live. Coke Fest was completely in line with the situation of COVID-19 in Pakistan and did not want to risk the event becoming a virus fiasco. The collaboration with FoodPanda was also completely digital. All the viewers had to do was "catch" deals being displayed on the side of the Tapmad website. They had to pay for the order and the food would be delivered to their home without any hassle. This event of Coke Fest has opened a new door for businesses to conduct their operations online.

The Internet is indeed a groundbreaking technology especially when it comes to developing countries but without developing a network of communications that following precedents designed by organizations; the knowledge sharing is nothing but raw and unfiltered data (Walther et al., 2005). Innovation without proper communication does not yield results as it should. This adds to the cost and does not reflect on the profits for any organization. Communication means sharing and receiving information from various resources (Cuevas-Vargas et al., 2020). Communications without managing knowledge go to waste. In other words, information technology gives room to develop a two-way communication stream that is easy to follow and helps in maintaining customer relations. In Pakistan, IT does have an established baseline in almost all industries, however, developing a two-way communication system still needs support.

Conclusion and Recommendations

According to Aziz (2020), Islamabad chambers of commerce and industry (ICCI) mentioned that PSL will be obliging in increasing the country's feeble economy by producing money and generating revenue through sports tourism as well as promoting a positive image of Pakistan. The businesses of airlines, textile industry, sports, advertisements, hotels and restaurants are actively participating in this league and gaining profits from it. Furthermore, the Pakistan cricket board (PCB) and PSL itself are building money via the auction of title sponsorship, broadcasting rights, ticket sales and much more. In addition, one of the most excellent things about PSL is that it provides recognition for the local talent on the world stage and their good performances will also lead towards higher ticket sales.

The effect of ICT on the economy is large, especially in developing countries where the need to expand businesses within a cost-effective frame is vital. Using the Internet to reach out to customers has helped Coke Fest establish and expand its fan base. It has reduced costs and increased revenues for the food industry. PSL on the other hand has boosted the economy by bringing in sports tourism across Pakistan which adds to the overall GDP. Coke Fest introduced Pakistan to the first digital event and carried it out successfully using an online ticketing system that enabled customers to login and view the event live. Coke Fest is the pioneer in making such a big event successful in Pakistan. Both events have also opened a new horizon on using social media and online ticketing systems to reach out to customers on a large scale. Having a completely digitalized method of ticketing can help both PSL and CF to smoothly manage their event and returns in case of a pandemic. Collaborating with various sponsors also has given room to PSL and Coke Fest organizers to increase overall profits, promote tourism and boost the economy at large.

Since most organizations are conducting their businesses online, it is recommended that they should move towards developing systems that support online marketing, tracking and ticketing. It is a cost-effective solution that will lower the costs incurred for printing tickets for any event. The location costs would be reduced as conducting an online/digital event will prove to be fruitful for the company and investors. New collaborations can be made with companies for conducting a digital event just like Coke Fest did with FoodPanda. Local companies and vendors can be promoted digitally at reduced costs. The customer reach would increase as it is easier to reach people via social media and websites. Website traffic can be increased organically.

References

Aftab, R., and Naveed, M. (2020). Investment review in sports leagues: financial evidence from Pakistan Super League. *Managerial Finance*, 47(6), 856–867.

Aftab, S., and Khan, M.M. (2019). Role of social media in promoting tourism in Pakistan. *Journal of Social Sciences and Humanities*, 58(1), 101–113.

Ahmed, Z., and Le, H.P. (2021). Linking Information Communication Technology, trade globalization index and CO_2 emissions: Evidence from advanced panel techniques. *Environmental Science and Pollution Research*, 28(7), 8770–8781.

Asghar, N., Batool, M., Farooq, F., and ur Rehman, H. (2020). COVID-19 pandemic and Pakistan economy: A preliminary survey. *Review of Economics and Development Studies*, 6(2), 447–459.

Aziz, H. (2020). PSL economic benefits. *The Express Tribune*. Retrieved from: https://tribune.com.pk/letter/2158822/6-psl-economic-benefits. (accessed 9 May 2021).

Bragg, M.A., Pageot, Y.K., Amico, A., Miller, A.N., Gasbarre, A., Rummo, P.E., et al. (2020). Fast food, beverage and snack brands on social media in the United States: An examination of marketing techniques utilized in 2000 brand posts. *Pediatric Obesity*, 15(5), e12606.

Briciu, V.A., and Briciu, A. (2021). Social Media and Organizational Communication. In M. Khosrow-Pour D.B.A. (Ed.), *Encyclopedia of Organizational Knowledge, Administration and Technology* (pp. 2609–2624). Hershey, PA: IGI Global.

Buhalis, D., and O'Connor, P. (2005). Information communication technology revolutionizing tourism. *Tourism Recreation Research*, 30(3), 7–16.

Cai, W., and McKenna, B. (2021). Knowledge creation in information technology and tourism research. *Journal of Travel Research*, 60(4), 912–915.

Cartwright, S., Davies, I., and Archer-Brown, C. (2021). Managing relationships on social media in business-to-business organisations. *Journal of Business Research*, 125, 120–134.

Cornell University, INSEAD and the World Intellectual Property Organization (WIPO) (2019). *Global Innovation Index 2019: Pakistan*. Retrieved from: www.wipo.int/edocs/pubdocs/en/wipo_pub_gii_2019/pk.pdf. (accessed 24 May 2021).

Cuevas-Vargas, H., Parga-Montoya, N., and Hernández-Castorena, O. (2020). Information and Communication Technologies to achieve an optimal relationship between supply chain management, innovation and performance. In J.L. García-Alcaraz, G.L. Jamil, L. Avelar-Sosa, and A.J.B. Peñalver (Eds.), *Handbook of Research on Industrial Applications for Improved Supply Chain Performance* (pp. 262–284). Hershey, PA: IGI Global.

Hong, Y., and Harwit, E. (2020). China's globalizing Internet: History, power and governance. *Chinese Journal of Communication*, 13(1), 1–7.

Ida, R., and Saud, M. (2020). An empirical analysis of social media usage, political learning and participation among youth: A comparative study of Indonesia and Pakistan. *Quality and Quantity*, 54(4), 1285–1297.

Invest Pakistan (2020). *Vision 2025, Annual 18–19 Plan and 11th Five Year Plan (2013–18)*. Retrieved from: https://invest.gov.pk/index.php/it-ites. (accessed 7 May 2021).

Iqbal, M.S., Rahim, Z. A., and Hussain, S.A. (2020). Industry 4.0 Revolution and challenges in developing countries: A case study on Pakistan. *Journal of Advanced Research in Business and Management Studies*, 21(1), 40–52.

Islamabad Policy Research Institute (2020). *Social Media Users in Pakistan*. Retrieved from: https://ipripak.org/ipri-infographics/. (accessed 24 May 2021).

Khan, A., Liaquat, S., Sheikh, J., and Pirzado, A.A. (2021b). The impact of Corona Virus (Covid-19) on e-business in Pakistan. *Journal of Contemporary Issues in Business and Government*, 27(3), 82–87.

Khan, S., Umer, R., Umer, S., and Naqvi, S. (2021a). Antecedents of trust in using social media for e-government services: an empirical study in Pakistan. *Technology in Society*, 64, 101400.

Lin, L., Shadiev, R., Hwang, W.Y., and Shen, S. (2020). From knowledge and skills to digital works: An application of design thinking in the information technology course. *Thinking Skills and Creativity*, 36, 100646.

Manjarrés, A., Pickin, S., Artaso, M.A., and Gibbons, E. (2021). AI4Eq: For a true global village not for global pillage. *IEEE Technology and Society Magazine*, 40(1), 31–45.

Memon, S., Mahar, S., Dhomeja, L.D., and Pirzado, F. (2015). Prospects and challenges for social media in Pakistan. In *2015 International Conference on Cyber Situational Awareness, Data Analytics and Assessment (CyberSA)* (pp. 1–5). IEEE, 8–9 June 2015.

Merriam-Webster (2021). *Social media*. Retrieved from: www.merriam-webster.com/dictionary/social%20media. (accessed 22 May 2021).

Müller, R., and Antoni, C.H. (2020). Individual perceptions of shared mental models of information and communication technology (ICT) and virtual team coordination and performance – The moderating role of flexibility in ICT use. *Group Dynamics: Theory, Research and Practice*, 24(3), 186–200.

Nizam, H.A., Zaman, K., Khan, K.B., Batool, R., Khurshid, M.A., Shoukry, A.M., et al. (2020). Achieving environmental sustainability through information technology: "Digital Pakistan" initiative for green development. *Environmental Science and Pollution Research*, 27, 10011–10026.

Organisation for Economic Co-operation and Development (OECD) (2020). *Connecting businesses and consumers during COVID-19: Trade in parcels.* Retrieved from: www.oecd.org/coronavirus/policy-responses/connecting-businesses-and-consumers-during-covid-19-trade-in-parcels-d18de131/ (accessed 22 May 2021).

Pentina, I., Koh, A.C., and Le, T.T. (2012). Adoption of social networks marketing by SMEs: Exploring the role of social influences and experience in technology acceptance. *International Journal of Internet Marketing and Advertising*, 7(1), 65–82.

Qalati, S.A., Yuan, L.W., Khan, M.A.S., and Anwar, F. (2021). A mediated model on the adoption of social media and SMEs' performance in developing countries. *Technology in Society*, 64, 101513.

Sattar, M.F., Khanum, S., Nawaz, A., Ashfaq, M.M., Khan, M.A., Jawad, M., et al. (2020). Covid-19 global, pandemic impact on world economy. *Technium Social Sciences Journal*, 11, 165–179.

Tirmizi, S.M. A., Malik, Q.A., and Hussain, S.S. (2020). Invention and open innovation processes and linkages: A conceptual framework. *Journal of Open Innovation: Technology, Market and Complexity*, 6(4), 159.

Walther, J.B., Gay, G., and Hancock, J.T. (2005). How do communication and technology researchers study the Internet? *Journal of Communication*, 55(3), 632–657.

World Trade Organization (WTO) (2020). *Export Prohibitions and Restrictions, Information note.* Retrieved from: www.wto.org/english/tratop_e/covid19_e/export_prohibitions_report_e.pdf. (accessed 23 April 2021).

Xinhua (2021). Pandemic-promoted online shopping becoming new normal in Pakistan. *The International News.* Retrieved from: www.thenews.com.pk/print/789606-pandemic-promoted-online-shopping-becoming-new-normal-in-pakistan (accessed 22 May 2021).

Zeb, Y. (2015). Pakistan ICT Report: Pakistan Is Still Ranked 112 According to Global IT Report 2015. *Research Snipers NEWS.* Retrieved from: www.researchsnipers.com/pakistan-ict-report-pakistan-still-ranked-112-according-global-report-2015/ (accessed 24 May 2021).

Zembylas, M., and Vrasidas, C. (2005). Globalization, information and communication technologies and the prospect of a "global village": Promises of inclusion or electronic colonization? *Journal of Curriculum Studies*, 37(1), 65–83.

7 Technology Application in Tourism Events

Reflections on a Case Study of a Local Food Festival in Thailand

Shirzad Mansouri

Introduction

Massive data and information is what tourists reflect and leave in the destination. Such huge data and information could be a kind of fundamental data that can pave the way for more development and efficiency in tourists' destination management. Fortunately, current information technology can manage and support massive amounts of data and useful content. One of the application areas that is becoming increasingly important in supporting corporate choices is business intelligence. Over the last few decades, academic and commercial communities have focused on business intelligence (BI) and business analytics (BA). Organizations view BI and BA differently, ranging from tools, techniques, technologies, and systems to strategies, processes, and applications that help enterprises make better and more timely decisions by analysing essential business data (Li et al., 2008).

Business intelligence concepts bring up opportunities to connect platforms to manage complex, unstructured data from a growing number of data sources, as well as to emphasize the analytical process of converting data into actionable strategies for better business decisions. Structures, databases, analytical tools, processes, and applications that assist people to more appropriate decisions are referred to as business intelligence (Turban et al., 2011). Data warehouse, business analytics, business performance management (BPM), and user interface are the four key components of business intelligence architecture (Turban et al., 2011). The foundations of BI are database administration and data warehousing, which are concerned with how data is gathered, structured, stored, extracted, and integrated so that end users may readily access it. The four main components of business intelligence architecture are data warehouse, business analytics, business performance management (BPM), and user interface (Turban et al., 2011). Database administration and data warehousing are the cornerstones of BI, and they deal with how data is obtained, formatted, stored, extracted, and integrated so that end users may easily access it.

Researchers such as Chen et al (2012) and Turban et al. (2011) believed that controlling, assessing and looking for similarities in a range of performance

DOI: 10.4324/9781003271147-10

measures defined as the basic pillars of a company tactics is the emphasis of business performance management (BPM).

In order for the business to produce new insights and better understand business performance (Chen et al., 2012), business analytics is defined as the wide use of data and quantitative analysis, which is typically based on data mining and statistical analysis. As a prominent firm in technology research, Gartner has identified four categories of analytics capabilities that may aid businesses in changing from classic descriptive analytics to advancing level of diagnostic analytics, predictive analytics, and prescriptive analytics (Rivera and Meulen, 2014). In diverse commercial applications, the techniques of data mining like designing decision trees, creation of neural networks, supporting vector machines, and cluster analysis are utilized for partitioning of the data, the prediction of the data modelling, analysis of correlation, cluster formation of data, and categorization as it was emphasized by Chen et al. (2012). A user interface (UI), commonly known as a visualizing of the data which is known as dashboard or data, allows mutual communication between two or more people.

A great range of measures such as business intelligence, customer relationship management to behavioural profiling, healthcare, and genetic analysis to supply chains, and analytics have been applied in various industries according to certain scholars like Davenport (2006) and Kusiak (2006). Further to this, Minelli, Chambers, and Dhiraj (2012) demonstrate how analytics of big data may be utilized to discover and answer business issues, allowing for competitive differentiation. Big data and analytics is described as the application of tactics for the data analysis to explain, investigate, and probe great and complex datasets that demand contemporary storing of data, as well as managing and visualizing technology.

Several studies on different dimensions of IB application in industry of tourism have been done. Meanwhile, Law, Leung, and Buhalis (2009) conducted a study of published papers on the rise of IT applications in tourism and hospitality research journals, which they classified into three categories: customers, technology, and suppliers.

Researchers such as Pyo, Uysal, and Chang (2002) discovered that data-mining techniques employed in destination management might be utilized to uncover information in data sets. The studies were conducted on certain topics such as clients, selling and buying locations, sales items and services, tourist destination advisors, and professionals in tourism industry. These studies cover everything from points in operation sections, various tools applied, and techniques to applications involving clients, participants, selling locations, purchasing items and services, destination advertisers, and tourism experts. A business intelligence approach was recommended in the study. They demonstrated how knowledge production, exchange, and application can all be done online using the example of online analytical processing (OLAP).

The case study looked at some of the contributions and effects of technology, such as business intelligence in travel, notably in event tourism. Mikuli, Paunovi, and Prebeac (2012), for example, utilized a model for networking and work-based importance-performance analysis (IPA) to determine visitor and exhibitor characteristics at a regional wine expo in Makarska, Croatia in their study.

The fact is that tourists and event participants consider professionalism and the event and exhibitions attraction to be the most essential components of a positive and favourable journey; moreover, the participants' experience will be mostly affected by the size of the exhibition and location choice of the event (Mikuli et al., 2012). Byrd and Gustke (2007), for example, presented a model of decision tree to consider the participation of beneficiaries in tourism planning, development of tourism, and its management, focusing on community views of tourism (Byrd and Gustke, 2007). Golmohammadi et al. (2011) introduced a rough and neutrally interconnected framework to predict tourists' general happy feeling with their travel experience in Iran and try to determine the significance of attributes that have an outstanding effect on their satisfaction, which affects their decision for choosing a destination and their willingness to come back to the event (Golmohammadi et al., 2011). The mentioned studies hardly consider the implications of using business intelligence and analytics insights in the tourism industry.

Vajirakachorn and Chongwatpol (2017) evaluated these studies and proposed a business intelligence model, as well as a business intelligence structure, to complete the BI implication in event tourism. Such a BI model for managing and converting data into valuable information in event tourism is shown in the following description. To aid analysts in deriving insight from visitor data, this model blends database administration, business analytics, business performance management, and data visualization architecture

The first stage is to establish the BI project's aims and objectives in the field of event tourism. Marketing analysts at event locations, for example, could be enthusiastic for building predictive models to aid management making decisions, with an emphasis on assessing the degree of overall satisfaction of the guests and visitors with their experiences during the trip or segmenting visitors using various criteria.

The second level of the business performance management (BPM) layer is to identify the key performance indicators connected to the strategy companies use. Some of these indicators are revenue, sales, incomes, expenses, scores obtained by satisfaction degree and their willingness to come back to the event location.

The next step entails gathering information important to the event tourism industry. For instance, it is critical to cover the attributes related to demographic features such as age, gender, income, and marital status in the business analytics process, as well as motivation-based features such as rest, health, activities, or experience and features of feeling worried in order to understand how participants choose their favourite location of travel such as security needed for giving safety, communication language, respective culture, or necessary hygiene.

The fourth stage, the level of the database management, is to combine data from various sources. This stage needs time since it involves data polishing, system formatting, and bringing data into organized order to prepare it for the perspective model development, particularly the time the analysts are dealing with issues like data inconsistency, duplication, and mistakes. The data visualization layer is created in the next step.

The fifth step is bringing more sense to the destination events and the people visiting the location. Exploring such data before model development allows the marketing experts to have a better understanding of how the destination's events and festivals are developed or the people that visit the site. To help with this kind of activity related to data exploration, Tableau Software creates a dashboard, the Data Visualization layer, which may be used on the Web or on a mobile device. This may assist analysts to first figure out what's going on with the main indicators explained in the second stage, and then comprehend tourists and event-related qualities.

The number of female and male participants depending on their age was reflected in three occasions during winter season, while the performance on the scene and activities during the event together all focused on the community and its social culture and their related background. The organizers obviously did not attract the attention of senior visitors in case the objective of this event is to include all participants from within and beyond the local context. Based on the post hoc study, the festival is not ideal for older guests due to uncomfortable facilities such as toilets and relaxing areas, as well as the location of the events, which are considerably further from the car park and require walking through the exhibition shops.

Within the sixth step, analysts examined the data to see if any odd patterns can be detected according to the first-hand analysis of information. For instance, the data experts may discover that about 50% of the visitors to the event location are from the north part of Thailand, and only 10% of them learned about the event by referring to the Facebook platform. On the contrary, just 20 guests come from the south side of Thailand yet more than 80% of them knew about the festival through the Facebook platform.

Analysts analysed the data in the sixth stage to determine if any unusual patterns could be discovered by analysing the data. For example, researchers discovered around half of the site's visitors are from northern Thailand, and only 10% of them knew about the event on Facebook. Guests from the southern part of the country make up just 20% of the total, but more than eighty percent of them learned about the event through the page of supporter of Facebook.

In the seventh stage, the business experts of analysis level are the matter of concern with building story cases on data analysis that may include a variety of techniques used for mining data like cluster analysis, logistic and polynomial regressions, decision trees, and neural network models.

These high-performance analytics, which are based on techniques of mining data, are employed and mixed to help analysts extract information from data. In particular, in the fifth and sixth stages, both data visualization and business analytics helped with the following questions on business analytics.

What happened? What was the reason for it to happen? What is going to happen, and how can we make it possible? It is also critical to choose a topic for your study that is both current and historical. During the high season such as school break and New Year's celebration, the event takes place eight weekends each year in the following months such as November, December, and January.

For example, "What are the significant factors that impact participants' willingness to return to the event location?" or "What kind of characteristics impact participants' happiness with what they experience at the event location?" Any data and important conclusions from exploring the data and its analysis during the fifth stage and sixth stage are presented in the eighth stage. The seventh stage is utilized to offer valuable input to the analysts in order to encourage any other marketing activities to keep tourists and other visitors for the coming events or practical strategies to increase the tourists and guests' pleasant experience. The ninth is to monitor and control such methods to guarantee that the supposing action parameters can be monitored and measured.

Research was carried out at "The 5th Walk to Remembrance at Naklua Market", a downtown festival in Pattaya, Thailand. On Pattaya-Naklua Street, there is an ancient fishing village called Naklua. The festival's organizers began the event with the goal of promoting Naklua's native cuisine and history. The event takes place over eight weekends in winter season every year. While walking in the old-style market area with wooden buildings on each side of the market, visitors may take in the local culture, entertainment, cuisine, and handicrafts. The festival's average number of attendees every evening is believed to be about 1000 people.

The information was gathered from around 3600 festival participants (150 surveys every night, three evenings a week for eight weeks). To prevent a tendency towards the location depending on the location of native participants, these samples are chosen using a stratified sampling approach. The questionnaires were delivered at the festival locations, and participants were free to participate in the survey.

Only 317 of the 2048 questionnaires (15.47%) were full and error-free, resulting in a response rate of 56.8% (2048) (data duplication and extreme values). All participants who responded were asked to assess opinions of the festival on a 5-point Likert scale starting from "strongly disagree" to "strongly agree", in addition to providing demographic information.

A total of 30 factors are used to explain and forecast the desire to return to event venues, including information related to the participants, push and pull factors, perceived values, perceived quality, and satisfaction.

The following results came out as a result of framework applied to the food festival:

Results from Step 1

The organizers of the festival search for solutions to not only delight event goers and urge them to return, but also to attract new guests through advertising, marketing, and positive word-of-mouth (WOM) from previous participants to their friends and family. As a result, the initial step in the BI model is to establish the project's goals.

Results from Step 2

Regarding the soaring number of events in each year, understanding visitor behaviour is more crucial than ever for festival organizers. Visitors come to events for a variety of reasons and with varied expectations, which influences their desire to go, their understanding of the value of the items purchased or the services quality they received, and thoughts they had. In the study, the researchers want to apply numerical models to know which dimensions of events and participants' perspectives have the most effect on their willingness to come back.

Results from Step 3

The collected information during the "Naklua Market" food festival in winter of 2014. The event drew over 24,000 people, with the largest attendance on the first and final days. The people in charge have access to information on the attendees, such as their age, sex, and salary. In order to optimize the favourable effects of the festival event, the significant winning components that affect the willingness to return to the event are assessed and indicated according to the degree of success their activities reached the expectations of participants. The event organizer's most significant responsibility is to realize the most important festival characteristics and present recommendations to the managing team in order to improve future festival experiences. The surveys gathered information based on parameters recognized as important in earlier research. Various studies have concentrated on the indicators of destination visitors' loyalty, as tourism destinations are mostly bound to repeated visits. Scholars such as Quintal and Polczynski (2010) consider how some factors such as festival attractions, quality of the event, related value, and security impact college students' satisfaction and, as a consequence, their academic success. According to Um, Chon, and Ro (2006), some independent variables that contribute to visitor loyalty include perceived attractiveness, perceived quality of service, perceived value for money, and contentment. Researchers discovered that the most important variable is perceived beauty, rather than absolute happiness (Um et al., 2006). Further to this, Jang and Feng (2007) investigated the impacts of visitors' attraction to its uniqueness and visitors' satisfaction about the event destination in order to understand temporal destination revisit intention (TDRI) measured on a short, mid and long-term basis.

Findings from Step 4

Since the obtained data emanated from various sources like form given for registration and questionnaires, the prior types of task in this step are to summarize and tabulate the data into the same framework, combine the obtained data for model development in Step 5, and delete any errors, as well as extreme values in the data. The main difficult element of data integration is dealing with missing values. To eliminate missing data, only completed replies are utilized to prepare the data for analysis. After removing replies with incomplete information, the dataset has 317 full observations that may be used to construct prediction models and cluster analysis to achieve the study's aim of finding characteristics that influence visitors' propensity to return.

Results from Step 5

The marketing analyst can acquire a feel of how the festival venue functions or how people engage with the site by exploring the data before building a model. Tableau Software produces a dashboard, the Data Visualization layer, to assist with such data exploration tasks. This dashboard may be utilized on the Web or on a mobile device. The dashboard assists analysts in establishing their preliminary results, which are discussed in the second phase. Festival activities are centred on the culture and background of the local community, as seen on stage shows. More insights are proven to be more beneficial for festival organizers in evaluating the market and creating a strategy for event preparations. One of the useful discoveries was that the number of ladies who attended outnumbered the men (more than 68% were women). Furthermore, it was discovered that over 55% of the visitors were between the ages of 20 and 39. Only 2% of the visitors were above the age of 60, according to the report.

In case the required objective of the festival is to engage all inside attendees as well as the outside ones, in and out of the social group, the organizers will not be able to succeed. According to post hoc research, the on-stage activities are not appealing, and the event is not pleasant for senior guests due to inconvenient amenities such as restrooms, rest spaces, and concession stands and parking area which were far for them to walk to the event arena. More was revealed when the majority of the visitors heard from their friends and relatives about the event. It means more serious marketing measures need to be taken and more re-evaluations needed for channels of advertising. More social media applications and channels need to be activated.

Results of Step 6

The results indicated a comparison between the average perceived value and perceived quality of items and services by visitors from both sexes throughout their stay. In addition, the data indicated that first-time male visitors under the age of 20 had a different opinion about activities done on the stage and general services than first-time visitors who are female and over 60.

Result from Step 7

In Step 7, scenarios were formed to analyse the data. Cluster analysis is used to figure out what visitors wish to receive out of their visit to the event location. The participants are classified into sub-groups based on their shared characteristics. Tourists partake in the festival for a variety of reasons: some may come to the festival to relax or investigate the grounds with his friends, while another may come to learn more about various cultures or socialize with new visitors on the event grounds; another individual might come to the event for the activities performed on the stage and the reasonably cheap food and drinks.

Predictive modelling is the second scenario. The dataset with its numerous predictor variables is analysed using four usual types of applications such as neural networks, stepwise logistic regression, decision trees, and support vector machines.

An artificial neural network (ANN), sometimes known as a network of human neurons, is a computational and computer framework that detects or categorizes patterns in data through a process of learning. The ANN model is a human-based inspired analytical method that simulates a body system like the brain's cells system, with a computer-based algorithm demonstrating how participants learn. ANNs have been effectively used in a number of sectors due to their outstanding ability to extract significant recognition of patterns or data categorization from data. The inaccurate data input, also known as independent variables, can be discrete or real-valued, whereas the results, also known as the targeted variable, are an accumulation of discrete or real-valued values.

The step-by-step technique focuses on selecting the factors known as predictors that best explain a specific predicted variable based on statistical criteria. A superior regression strategy called stepwise polynomial logistic regression is used to predict a dependent variable using "n" independent variables. When the link between the goal variables and the explanatory variables is intricate and non-linear, it is widely utilized.

One of the best supervised learning algorithms is SVMs (Support Vector Machines), which are based on the concept of decision planes setting decision boundaries. SVMs use the mapping function to generate hyperplanes in a multi-dimensional space in order to categorize data or predict the numerical value of the desired output in a resample.

A decision tree is a well-known data categorization and prediction technique due to its simple explainability properties. A decision tree divides a dataset into multiple groups by evaluating individual data items based on their attributes. Due to its straightforward explainability qualities, a decision tree is a well-known data categorization and prediction approach. By analysing individual data items based on their qualities, a decision tree separates a dataset into multiple groups.

The major findings aid festival organizers in prioritizing the critical aspects that influence the likelihood of subsequent visits to the event site. Given enough data and the right input pieces, advanced business analytics using data mining techniques (cluster analysis and predictive modelling) may be able to understand visitor profiles, according to this study.

Implications for Managers and their Practices

The festival organizers clearly could not attract the attention of senior citizen participants if the objective of this event is to include all visitors from within and beyond the local community. According to post hoc study, the festival's on-stage and activity programming are unappealing, and inconvenient amenities like restrooms, rest places, and food booths make it unsuitable for elderly guests.

Several tourism-related apps make extensive use of its capabilities. The current study stresses the value that tourism planners, organizers of events, and researchers in marketing at festival locations can provide through the use of big data, business intelligence, and analytics, which have helped top-performing companies deliver their products and services successfully.

Traditional statistical analysis methods such as spreadsheet-based regression modelling, cross-tab analysis, and descriptive statistics also lack advanced predictive analytics, which are clearly required to analyse large amounts of data in order to understand participant satisfaction with their trip experiences and make appropriate decisions.

In this study, the business intelligence tool recommended for gaining insight into the behaviour of people visiting tourism destinations stands out. Festival organizers, in particular, must prepare carefully and analyse data and information gathered from both returning and non-returning attendees.

The capacity to use a data mining approach to uncover and analyse participant characteristics has become a requirement for attracting and retaining the event's most significant visitors. Despite the fact that this study was conducted in conjunction with a small Thai event, it shows how tough it is to attract repeat visitors to locations or festivals in a competitive market.

This model can be used to look into the practical applications of business intelligence and analytics in any tourism industry, expanding knowledge and practice.

How can we create an effective BI solution to help us figure out why people want to come back to the event or not? To meet the festival organizers' expectations and needs, the first challenge is to create acceptable goals and objectives, as well as to estimate the project timeline while keeping within budget. Unreasonably high expectations always diminish perceived value.

This model can be used to look into the practical applications of business intelligence and analytics in any tourism industry, expanding knowledge and practice.

The second obstacle for business intelligence projects is data integration for model generation in Step 4. As a result, 60% of the project time is spent preparing, integrating, and cleaning data to ensure its quality before applying data analytics. The next challenge is Step 7: Creating a Data Analysis Framework, which requires a shift in focus from description to analysis.

Analysts in marketing and event organizers chose to skip Steps 7 and 8 of the BI architecture because they are more skilled with data visualization and inquiry in Steps 5 and 6. As a result, marketing analysts and event planners will be able to experiment with analytical tools in order to improve their results.

The second obstacle for BI projects is data integration for model generation in Step 4. As a result, 60% of the project time is spent preparing, integrating, and cleaning data to ensure its quality before applying data analytics. The next challenge is Step 7: Creating a Data Analysis Framework, which requires a shift in focus from description to analysis.

Analysts in marketing and event organizers chose to skip Steps 7 and 8 of the BI architecture because they are more skilled with data visualization and inquiry in Steps 5 and 6. As a result, marketing analysts and event planners will be able to experiment with analytical tools in order to improve their results.

Final Reflection

This chapter's case study is based on a real-life event tourism application. Event organizers are looking for ways to satisfy festival attendees and entice them to return, as well as to attract new visitors through advertising, marketing, and positive word-of-mouth from existing attendees to their family and friends.

This business intelligence tool might be used to provide vital feedback to event organizers, allowing them to focus on any follow-up marketing efforts to retain attendees for future events, as well as effective approaches to improve participant satisfaction.

Festivals and events that seek to acquire tourist authority support or partnership, or just earn greater respect, conduct tourism and economic impact studies to "prove" their economic significance. Their strategy may be to establish oneself as a tourist attraction first, then use that position to gain reputation and encourage expansion. According to stakeholder and resource dependence theory, in order to be sustainable, events must gain physical resources and political backing, losing some autonomy in the process. A business intelligence application might be a way to get additional political and physical resources, which would contribute to long-term sustainability.

Various studies and works focusing on the application of business intelligence in other forms of event and festival tourism in Thailand appear more plausible. This will include business and tourist events, sporting events and tourism, festivals, other cultural festivities, the Olympics, world fairs, and other major events.

Furthermore, researchers must explore the use of theory and methodologies from linguistics, psychology, anthropology, and sociology, as well as their integration with business intelligence, in order to ease the decision-making process for planning and managing event tourism.

References

Byrd, E.T., and Gustke, L. (2007). Using decision trees to identify tourism stakeholders: The case of two eastern North Carolina counties. *Tourism and Hospitality Research*, 7(3–4), 176–193.

Chen, H., Chiang, R.H.L., and Storey, V.C. (2012). Business intelligence and analytics: From big data to big impact. *MIS Quarterly*, 36(4), 1165–1189.

Davenport, T.H. (2006). Competing on analytics. *Harvard Business Review*, 84(1), 98–107.

Golmohammadi, A., Shams Ghareneh, N., Keramati, A., and Jahandideh, B. (2011). Importance analysis of travel attributes using a rough set-based neural network. *Journal of Hospitality and Tourism Technology*, 2(2), 155–171.

Jang, S., and Feng, R. (2007). Temporal destination revisit intention: The effects of novelty seeking and satisfaction. *Tourism Management*, 28(2), 580–590.

Kusiak, A. (2006). Data mining: Manufacturing and service applications. *International Journal of Production Research*, 44(18–19), 4175–4191.

Law, R., Leung, R., and Buhalis, D. (2009). Information technology applications in hospitality and tourism: A review of publications from 2005 to 2007. *Journal of Travel & Tourism Marketing*, 26(5–6), 599–623.

Li, X.R., Cheng, C.K., Kim, H., and Petrick, J.F. (2008). A systematic comparison of first-time and repeat visitors via a two-phase online survey. *Tourism Management*, 29(2), 278–295.

Mikulić, J., Paunović, Z., and Prebežac, D. (2012). An extended neural network-based importance-performance analysis for enhancing wine fair experience. *Journal of Travel & Tourism Marketing*, 29(8), 744–759.

Minelli, M., Chambers, M., and Dhiraj, A. (2012). *Big Data, Big Analytics: Emerging Business Intelligence and Analytic Trends for Today's Businesses*. Hoboken, NJ: John Wiley & Sons.

Pyo, S., Uysal, M., and Chang, H. (2002). Knowledge discovery in database for tourist destinations. *Journal of Travel Research*, 40(4), 374–384.

Quintal, V.A., and Polczynski, A. (2010). Factors influencing tourists' revisit intentions. *Asia Pacific Journal of Marketing and Logistics*, 22(4), 554–578.

Rivera, J., and Meulen, R. (2014). Gartner says advanced analytics is a top business priority. *Gartner*. Retrieved from: www.gartner.com/newsroom/id/2881218

Turban, E., Sharda, R., and Delen, D. (2011). *Decision Support and Business Intelligence Systems*. Upper Saddle River, NJ: Pearson Education.

Um, S., Chon, K., and Ro, Y. (2006). Antecedents of revisit intention. *Annals of Tourism Research*, 33(4), 1141–1158.

Vajirakachorn, T., and Chongwatpol, J. (2017). Application of business intelligence in the tourism industry: A case study of a local food festival in Thailand. *Tourism Management Perspectives*, 23, 75–86.

8 An Assessment on Strengthening the Attractiveness of Turkey's Event and Festival Tourism with Innovative Technology Efforts

Kaplan Uğurlu

Introduction

Events such as festivals, fairs, congresses, meetings, celebrations, ceremonies, fairs, feasts, and so on, are an indispensable element of culture and tourism, organized or participating in events at the regional, national, or international level. The event can also be defined as the systematic planning, development and marketing of a number of activities that revitalize the natural and physical resources located in the city in which it is held, create an image for that place and as a tourist attraction (Getz, 1997).

The event provides income and employment opportunities to the destination, increasing the quality of life for local people and causing new investments to be made in the region and the infrastructure to be renewed. The events such as festivals are organized to showcase the cultural diversity of the region, to promote the settlement, to increase entertainment opportunities for local people, to attract visitors to the region, thus, to improve the quality of life in the place where they live and to create income. Event tourism is an important type of tourism that is growing rapidly in the world tourism mobility. Events are an important motivating factor in tourism. Based on this importance of event tourism, many destinations are trying to benefit by organizing various events. The Olympic Games, Formula 1, FIFA World Cup, UEFA Champions League can be given as examples of sports events held around the world. Examples of cultural, artistic and entertainment events are the Eurovision Song Contest, Oscar Award Ceremony, Grammy Award Ceremony, Cannes Film Festival, Rio Carnival, and so on.

About 1400 small and large events are held annually in many regions of Turkey to keep the rich Anatolian culture alive, improve the image of cities, promote urban development, stimulate economic, cultural, and social life, attract visitors and investments (Kültür Portalı, n.d.). Lack of organization, finance, and promotion of events and festivals in Turkey result in long-winded problems and local managers, local communities, regulatory agencies, and organizations short- and long-term failed to meet expectations (Yolal, 2017). With the COVID-19 pandemic, businesses have started to carry out their activities remotely and virtually through online and mobile tools as much as possible within the framework of

DOI: 10.4324/9781003271147-11

pandemic rules. Classical marketing activities have been digitized and contemporary marketing approaches have started to be implemented by businesses as technology intensive. These changes in lifestyle, business and markets include events and festivals. The use of new technologies in events and festivals is now vital. The sustainability, profitability, awareness and competitiveness of events and festivals are directly proportional to their compatibility with technology. In this study, events and festivals will be discussed comprehensively, the usability of technology in events and festivals and its contributions to Turkish tourism will be evaluated.

Definition, Scope and History of Events and Festivals

Getz (2008), who stated that events are a subject that should be investigated in terms of marketing destinations, defined events as remarkable activities that occur in a certain place and time, under certain conditions, that there is an element of attractiveness for a destination (Getz, 2007). Goldblatt (2008) draws attention to the celebration aspect of the events and defines the events as having a unique time with ceremonies and rituals organized to meet special needs. From the definitions, it can be said that events are an element of attraction for destinations, an activity in which people visiting a destination participate in a certain place and at their leisure and in certain conditions, have their own ceremonies and rituals, and are remarkable activities. In order to participate in the activities, people travel to the country or city where the events take place, spend time there, stay, eat and drink, have fun and travel, making the activities a marketing tool that attracts tourists and visitors to the region.

The concepts of event and festival are different from each other. Although there are conflicts, it is possible to divide activities into four broad categories based on whether they have cultural, leisure, personal and organizational goals (Shone and Parry, 2004). A festival will be covered by both leisure and culture. Festivals are different from events. Festivals are events that involve various activities and in which people participate for different purposes. In essence, festivals are celebrations that offer a chance to finalize ordinary daily life (Anderton, 2008). According to Quinn (2013), festivals are social practices that communities have been engaged in for generations with the purpose of expressing and celebrating their beliefs. Quinn (2013), on the other hand, also acknowledges that there is some confusion due to the complications of determining which event can be categorized as a festival. Although different definitions have been made in different disciplines related to festivals, festivals are events consisting of special activities such as culture, art and sports that are used in the image of modern tourism. Festivals are events that serve to strengthen the ties between society, culture, and the environment, which have become part of people and society (Goldblatt, 2000).

According to Getz (1997), events have functions such as marketing the destination, creating tourist attractions, creating an image, supporting existing values and encouraging tourist movements to the destination. Events in destinations will attract local, national, and international tourists as well as investors who will invest in the region. Thanks to the positive image that the events will bring to the region, both tourists will have a pleasant experience and the economic vitality that tourists bring to the region will increase the economic well-being and quality of life of the local people. The beginning of event tourism in the destination will be an occasion for more people to come to the region, spend time in the region, stay overnight in the region, use food and beverage services and make various purchases, thus creating an alternative tourism demand for the region. The destinations where the events are held will contribute to the erasure of the bad image, if any, while gaining a new image. Due to the fact that the event zone is a tourist attraction area, the region will have taken a serious step towards becoming a brand city that will acquire a new urban identity with the sub-investments made by local and central governments in the region.

It is assumed that the events date back to very ancient times (Çoban, 2016). It is known that planned events were held as early as 476 BC in the period before the fall of the Western Roman Empire (Raj et al., 2013). Various historians consider the first festivals to have occurred in Ancient Egypt about 4500 years ago, which are considered to include religious ceremonies and political festivals, as well as music and dance (Curran, 2018). Nowadays, various celebrations, anniversaries, weddings, ceremonies, festivals, fairs, and so on, continue to be held in modern societies.

Development and Necessity of Innovative Technologies in Events and Festivals

Technology continues to develop rapidly and dominate our lives. Nowadays, enterprises that do not adapt to technology have almost no chance to compete in the market. The profitability, sustainability and social usefulness of enterprises depend on the satisfaction of consumers. It is important for businesses to anticipate this demand by looking for opportunities to provide differentiated, fast, easy-to-use, and stable products and services to improve consumers' lives. In order to provide a unique experience to the participants, event and festival professionals have also started to make great use of technological innovations in their operations. The time, venue, cost, organizer, target audience and program of events and festivals are a specific project. Even at the smallest event, income and expense calculations, determination of the number of participants, ticketing, payment systems, heat, light and sound arrangements, animation, video and demonstration systems, security, and inventory tracking, and so on, technology is used.

The most important thing to pay attention to when investing in technology at events and festivals is that the technological investment to be made is not higher than the profitability that the event will bring. In addition, it should be considered that the technology investment should save time and money, and with the use of communication and information technologies, you should be able to quickly and easily communicate and exchange information with customers and business partners. It should be ensured that the features and functions of the technology used are working smoothly to eliminate technical errors and malfunctions that may occur during the event. Your knowledgeable and equipped staff who can use technical machines, tools and devices should be employed throughout the event. Technology should make less use of human power. The technology used should ensure that the participants have a fun and pleasant time during the event. Also, technological innovations should be used professionally for the promotion and marketing of the event.

The invention of the Internet is undoubtedly as important as the invention of electricity. The Internet, computer and intelligent software technologies, communication and information technology, social media and networks have become an integral part of everyday life. With the COVID-19 pandemic, education, shopping, meetings, fairs, virtual tours, virtual museum visits, virtual libraries, and so on, have started to be held virtually over the Internet. This COVID-19 pandemic has proved to us that events can be held virtually thanks to technological innovations. There is no longer any doubt that global technological innovations will continue to develop, new technologies will be used not only in technology-intensive industries, but also in the service sector and events. In other words, in order to create a successful community of people and to have a social and cultural impact on the environment, it will be enough to have a computer, smart mobile phone, and so on, connected to the Internet and mobile networks, rather than coming together in a venue anymore. A brief example of some of the changes that have occurred in technology in the event industry of Goldblatt (2014) is shown in Table 8.1.

It is now much more important for event managers to stay up to date with world technologies and trends in changing market conditions. Event organizers are now looking for new solutions to make their lives easier and provide an enriching experience to their participants. Event planners, on the one hand, try to influence the participants, keep their perceptions and thoughts positive and give them a unique event experience, while on the other hand, they strive to increase ROI (return on investment) they make.

Table 8.1 Paradigmatic changes in events and festivals

FROM	TO
Analog	Digital
Collision	Collaboration
Content	Context
Event	Events without end
Live	Blended
Local	Cloud
Hardware	Software
Human Staff	Technological Stuff

Source: Goldblatt, 2014

As mentioned, technology is changing our communication processes and making online participation possible. This is due to the emergence of social media and mobile communications. For example, participants in the event share their event experiences with an instant message, image, or video using social media as a normal way to connect. In earlier times, the point of engagement was only one – the event itself. The tedious and outdated processes have been working for quite some time, and some companies still use them, but as people have become more tech savvy, expectations have also grown that these traditional methods will no longer be able to cope.

Event managers conduct the events easier, faster, and successfully before, during and after the event (e.g., planning, workflow, registration, ticketing, tracking, etc.) with new event software and real-time connections together with event stakeholders (managers, organizers, sponsors, employees, volunteers, participants, etc.). Some of the technologies that can be used to make events effective and successful are as follows (Sitepara, 2017): Internet marketing, mobile marketing, vent application, drone, network platform, QR Code/RFID (Radio Frequency Identification), Beacons, the Audience Response System (ARS), Wi-Fi, virtual events/meetings. Global companies prefer virtual events due to advantages such as reducing travel costs, reducing out-of-office time, allowing participants to communicate, holding a meeting that will be cancelled due to budget, and sharing information in multiple markets.

An Assessment of Innovative Technologies to Strengthen Turkey's Event Tourism

As with any nationality, when we look at the history of the Turks starting from Central Asia and extending to the present day, nature, climate, lifestyle, traditions, faith, migrations, war, agriculture, and so on, it is known that they are engaged in activities related to human life and culture. Religious and national holidays, celebration of the arrival of spring, highland festivals, weddings, animal trading market events, harvest abundance and fertility prayer events are just some of the events still going on from very ancient times to the present day. To give some examples of events from Turkish history, during the Göktürk period the rite of sacrifice for Tengri, which is a sacred place and water, and the sacrifice of an animal for Allah is still celebrated as a holiday. For Uyghur Turks who lived in hot climates in history, a cold food festival was also held, and people at the festival were entertained by soaking each other. Examples of other events that have come down to the present day include camel wrestling, oil wrestling competitions, horse races, and so on.

While the festival culture in Turkey began to be observed in the 1960s with the name of the festival, the process of spreading the festivals throughout the country coincides with the years after 1980 (Akarpınar, 2004). A significant part of local, national, and international festivals in Turkey are associated with agricultural products. These festivals, which are mainly agricultural, are usually referred to by the name of vegetables, fruits, grain products that are identified with the geographical region where they are grown. The content of these festivals is often enriched, and their scope can also be expanded in the form of culture, culture-art, culture-tourism, culture-industry, industry-trade. In geographical regions where livestock activities are intense in Turkey, festivals are organized where livestock and animal products come to the forefront. In addition, it is organized at artistic and cultural festivals such as Jazz, Opera, Ballet, film festivals. It is possible to say that two very important needs of the Turkish society come to the fore when evaluating festivals in Turkey. First, that festivals retain their importance in terms of exhibiting local, regional, economic, cultural values and development; second, that they are instrumental in gaining universal values and integrating with the world. In Turkey, some well-known festivals are as follows: Hitchhiking Festival (Bursa), Zeytinli Rock Festival (Balıkesir), Thrace Music Fest (Tekirdağ), Bozcaada Jazz Festival (Çanakkale), Aspendos International Opera and Ballet Festival (Antalya), Fest Million (Istanbul), Adana Film Festival (Adana), The Golden Orange Film Festival (Antalya), Kirkpinar Oil Wrestling (Edirne), Thrace Vineyard And Ecology Festival (Tekirdağ), Alacati Herbs Festival (Izmir), Polonezköy Cherry Festival (Istanbul), Kakava-Hidirellez Festivals (Edirne), the Rose Harvest Festival (Isparta), the Mesir Macunu Festival of Manisa (Manisa), and Adana Taste Festival (Adana). The data of the festivals organized in Turkey according to their themes are shown in Table 8.2.

The sustainability of the events depends on the satisfaction of the visitors attending the event and the event being memorable for everyone. Event organizers can provide participants with a unique experience thanks to technological innovations before, during, and after the event, ensuring that activities such as culture, art, business, entertainment, ceremonies, meetings, and so on, with heat, light, sound, and visual technologies will increase the quality of the

Table 8.2 Distribution of festivals in Turkey by theme in 2019

THEME	NUMBER	RATE (%)
Culture and Art	261	28.43
Agriculture	238	25.92
Culture	88	9.58
Sport	72	7.84
Gastronomy	63	6.86
Nature	60	6.53
Culture and Tourism	58	6.31
Children and Youth	31	3.37
Season	16	1.74
Mining	13	1.41
Economic	5	0.54
Ecological	4	0.43
Other	9	0.98
Total	918	100

Source: Demir, 2019

event, increase the motivation of the participants, sponsors, volunteers, and employees, while ensuring that the event activities are carried out in a planned and organized manner. With a greater efficiency, technology will be of great benefit in solving complex activities. Technologies are used seriously in international festivals and fairs held in Turkey. For example, since it was selected as the 2010 European Capital of Culture, the central government, Istanbul Municipality, non-governmental organizations, and tourism companies acted together for the promotion of Istanbul, the success and profitability of the events with e-marketing technologies (İnan and Ölçer, 2010). First of all, Istanbul 2010 European Capital of Culture Agency was established in order to coordinate the work of public institutions and non-governmental organizations in order to prepare them for the events that took place in 2010. In particular, the 2010 European Capital of Culture website (https://istanbul2010.org/) was created, promotions were focused on the Internet and social media, then the features of the website, Google page rankings and feedback link rates were examined. The objectives were achieved through interviews. Cultural and artistic activities such as music, theatre, exhibitions, panels, symposiums, meetings, fairs, and festivals were not only included in the events, but also many historical and cultural properties were restored in 2010. The events have been very successful.

The most influential institutions in the marketing of events are the relevant Ministries of Governments, and then other public and private institutions in the tourism sector. Ministry of Culture and Tourism, Turkish Hoteliers Union, Travel Agencies Association, Tourist Guides Association, and leading enterprises in the tourism industry provide content-rich information about Turkey's cultural resources, places of interest and planned events. Among these contents are the national and international visual arts, music and opera, concerts, films, documentaries, animations, literature, theatre, performing arts, traditional arts, urban and rural culture, education, cultural heritage, museums, and sports activities (Okumuş et al., 2012). Such detailed information can help and motivate domestic and foreign tourists and visitors who want to participate in events to visit and prepare for activities related to participation in event tourism.

Technologies are used in events in Turkey, and the Teknofest Festival on the theme of aviation, technology and space technology is the first and only festival aimed at developing Turkey's national technologies. In general, it aims to promote and develop national technologies and raise people's awareness in this field. The festival includes seminars, award-winning technology competitions, domestic technology exhibitions, international entrepreneurship summit; skydiving, Solo Turkish, Turkish Stars (acrobatic flights), and similar local and international air shows are also held. The festival is held one year in Istanbul and one year in one of the Anatolian provinces (Teknofest, n.d.). In total 1,750,000 people visited Teknofest Istanbul Festival (Habertürk, 2019). Science festivals of different sizes are also organized by universities and industries, such as Researchers' Night Event, Euroscience Open Forum, National Patent Fair and University-Industry Cooperation Congress, and TUBITAK National Sky Observation Festival.

With the development of the Internet and mobile technologies, e-commerce is becoming increasingly widespread, and visitors who demand events and festivals can get information about events, attend, or watch events remotely, make reservations for events purchasing tickets online and make payments through these technologies. Technology also causes benefits such as effective communication, navigation, location finding. It is possible to send letters, pictures, audio, and video images to any place whose address is known, without time limit, via e-mail, which is one of the important areas of use of the Internet. In the promotion of the events, banner ads and advertisements can be made in a coordinated and versatile manner with all Internet-based and mobile tools. The event website can have rich content and functionality. Promotional films, photo galleries, virtual tours, informative maps, animations, promotional CDs, e-books, e-magazines, e-bulletins, e-brochures, e-newspapers, news, maps, announcements, promotions, statistics, transportation, weather and climate, local time, exchange rates, e-tools, e-mails, request-suggestion-complaint form, surveys, frequently asked questions, guestbook, forum, advice, information, visitor counter, site search, site map, online reservation, customer services, human resources, foreign language and links constitute the elements that can be made available to the visitor on promotional sites. Based on the increasing importance of social media recently, the fact that entries to these sites are linked to social media memberships will both increase their accessibility via social media and increase the effectiveness of the site. For example, an integrated digital promotion campaign "Web, Mobile and Social Media" was conducted on 10 different platforms, 362,000 loyal followers' members from 130 countries, tens of thousands of publications in 27 countries by Boom, on tens of thousands of publications for Istanbul Shopping Fest, which was held by the General Directorate of Promotion in 2011 (Çakır and Yalçin, 2012).

Conclusion and Recommendations

Events are a new tourism product that will create demand for destinations. In addition to the economic, socio-cultural, and environmental impacts of events on a destination, events are also an attractive strength for destinations that event venues, cities and countries want to create a new event or host an existing event. There can be many reasons why a destination might make an event as an attraction. These are to attract more visitors to the region, to ensure the economic development of the local people, to increase the image and recognition of the region, to encourage investments to be made in the region and to improve the infrastructure of the destination, and so on.

People participate in activities for different purposes and want their needs and desires to be satisfied. There are different motives that motivate participants to join activities. These are entertainment and relaxation, participation in cultural and artistic activities, being together with family, friends or a community, and for prestige and status. People who come to the destination to participate in the event also consume different tourism products. The expenses incurred by event consumers for accommodation, travel, food and beverage, entertainment, and similar needs and wishes are much more than the money paid for transportation and tickets for participation in the event. That is why it is so important that the events are planned and successfully perform their functions. As much as the sustainability and profitability of the event, it is necessary to ensure that consumers of the event participate in the events again, even those who have never participated in the event before. Determining the impact of events on this process and planning marketing activities in this direction can also contribute to destination marketing.

With the COVID-19 pandemic, the need for businesses to benefit from technological innovations much more has emerged in whichever sector they are in. Events and festivals are among the sectors that are keeping up with technology. With digitalization, the expectations of the event and festival participants change from the experiences presented to them, and these expectations are concentrated on four main themes (TURSAB, n.d.). The first of these is personalized service. By analysing the demographic variables of the people participating in the event, the target audience of the participants can be determined, and the product and service mix can be designed accordingly. Second, be inspired and inspiring. With the rise of the Internet, mobile and social media, comments, photos, blog posts, and so on, have started to play an important role in the research, decision-making and participation processes of the participants in the event. Now the total image of event is formed not only within the framework of its own creation, but also through the sharing of feelings and thoughts by the participants. In this direction, it becomes important to manage the perception created by the visitors participating in the event as a result of their own experiences. Third, being mobile. As of 2021, the number of singular mobile users has reached 77 million in Turkey and 5.2 billion in the world (We Are Social, 2021). In a world where mobile usage is increasing day by day, mobile and continuous Internet access for participants is becoming an indispensable part of the event experience. Mobile-enabled experiences are included in the most important expectations of the participants at every stage of the event experience. Internet and mobile-based communication and information technologies will make easier the work of both participants and event organizers. Fourth, easy and quick solutions. Event managers and participants want to achieve results with practical solutions in their end-to-end experience. At the events, participants highlight time-saving experiences such as voice or visual search, self-service tablets, one-button payment. In order to meet the expectations of the participants and to stand out in the competition, human-oriented approaches instead of product-oriented approaches are becoming more important every day. The problem that has been achieved with a human-oriented is not only the participants, but also the answer to business problems such as competition, loss of market share, cost faced by organizers and managers, and in return, it returns as an increase in performance on metrics such as growth, profitability, and participant loyalty.

Turkey has an important potential in terms of event tourism with its natural, historical, and cultural resources. However, it does not seem possible to say that this potential can be used effectively and efficiently. The main purpose of an event is to provide a celebration, commemoration, and entertainment opportunity for the public. Proceeding from this, the chances of a successful event that does not involve residents are quite low. Therefore, in order to increase the expected benefits from organizing events, reduce the negative effects and turn the event into a long-term form of celebration, it is also necessary to plan, organize, and manage events strategically. In order to achieve the success of the event, modern technologies should be used to create awareness about the quality of the events. Compliance with technology is being achieved rapidly in Turkey and we can see this at the events held. However, the high costs imposed by technology should be reduced and local people should not be removed from events due to the high prices. Otherwise, the events will go no further than the events attended only by a wealthy group. The events provide public benefits to the destination, the city and the country in terms of development and promotion. For this reason, all public and private event stakeholders should make effective use of communication and information technologies, social media, and e-marketing methods. It should be remembered that events and destinations market each other.

References

Akarpınar, B. (2004). Tarım toplumundan sanayi toplumuna geçişte panayır-sergi-fuar-festivallerin durumu ve Türkiye örneği. *Milli Folklor Dergisi*, S. 64, pp. 25–36.

Anderton, C., (2008). Commercializing the carnivalesque: The V Festival and image/risk management. *Event Management*, 12(1), 39–51.

Çakır, M., and Yalçin, A.E. (2012). Kültür ve turizm tanıtımında bir araç olarak internet kullanımı. Retrieved from: https://teftis.ktb.gov.tr/Eklenti/4715,kultur-ve-turizm-bakanliginda-bir-arac-olarak-internet-.pdf?1 (accessed 8 August 2021).

Çoban, Ö. (2016). Event tourism. In C. Avcikurt, M. S. Dinu, N. Hacioglu, R. Efe, A. Soykan, and T. Nuray (Eds.), *Global Issues and Trends in Tourism*. Sofia: St. Kliment Ohridski University Press, pp. 459–470.

Curran, V. (2018). *An investigation to identify how festivals promotional techniques have developed over the years- Using Green Man Festival as a case study*. Retrieved from: https://repository.cardiffmet.ac.uk/bitstream/handle/10369/10039/Dissertation%20-%20Victoria%20Curran.pdf?sequence=1&isAllowed=y (accessed 9 August 2021).

Demir, S. (2019). Coğrafi açıdan Türkiye'de festivaller. *Unpublished Master Thesis*. Isparta: Süleyman Demirel Üniversitesi Sosyal Bilimler Enstitüsü Coğrafya Anabilim Dali.

Getz, D. (1997). *Event Management and Event Tourism*. New York: Cognizant Communication Corporation.

Getz, D. (2007). *Event Studies: Theory, Research and Policy for Planned Events*. Oxford: Elsevier.

Getz, D. (2008). Event tourism: Definition, evolution, and research. *Tourism Management*, 29, 403–428.

Goldblatt, J. (2000). A Future for event management: The analysis of major trends impacting the emerging profession. In J. Allen, R. Harris, L.K. Jago, and A. J. Veal (Eds.), *Events Beyond 2000: Setting the Agenda Proceedings of Conference*. Australian Centre for Event Management, pp. 2–9.

Goldblatt, J. (2008). *Special Events: Event Leadership for a New World.* Hoboken, NJ: John Wiley & Sons.

Goldblatt, J. (2014). *International Centre for the Study of Planned Events.* Edinburgh: Queen Margaret University.

Habertürk (2019). *TEKNOFEST İstanbul'u 1 milyon 750 bin kişi gezdi.* Retrieved from: www.haberturk.com/teknofest-istanbul-u-1-milyon-750-bin-kisi-gezdi-2524587-ekonomi (accessed 6 October 2021).

İnan, E., and Ölçer, C. (2010). Büyük ölçekli etkinlik yönetiminde e-pazarlama çalışmaları: İstanbul Avrupa Kültür Başkenti örneği. *Erciyes İletişim Dergisi,* 1(4), 7–15.

Kültür Portalı (n.d.). *Etkinlikler.* Retrieved from: www.kulturportali.gov.tr (accessed 6 October 2021).

Okumuş, F., Avcı, U., Kılıç, İ., and Walls, A.R. (2012). Cultural tourism in Turkey: A missed opportunity. *Journal of Hospitality Marketing & Management,* 21(6), 638–658.

Quinn, B. (2013) Arts festivals, tourism, cities, urban policy. In D. Stevenson and A. Matthews (Eds.). *Culture and the City: Creativity, Tourism, Leisure.* Abingdon: Routledge, pp. 69–84.

Raj, R., Walters, P., and Rashid, T. (2013). *Events Managements Principle and Practice.* London: Sage.

Shone, A., and Parry, B. (2004). *Successful Event Management: A Practical Handbook.* London: Thomson.

Sitepara, J. (2017). 8 event technologies every organiser must embrace. *Medium.* Retrieved from: https://medium.com/hubilo-official-blog/8-technologies-all-event-planners-must-embrace-2910b1371ebf (accessed 9 August 2021).

Teknofest (n.d.). *Sıkça sorulan sorular.* Retrieved from: www.teknofest.org/sss.html (accessed 18 September 2021).

TURSAB (n.d.). *Turizm sektörü dijitalleşme yol haritası.* Retrieved from: www.tursab.org.tr/apps//Files/Content/ad5f3ddb-5a11-410f-9e3c-fe8b2dc4df8b.pdf (accessed 9 August 2021).

We Are Social (2021). Digital around the world. *Dijilopedi.* Retrieved from: https://dijilopedi.com/2021-dunya-internet-sosyal-medya-ve-mobil-kullanim-istatistikleri/ (accessed 8 August 2021).

Yolal, M. (2017). Türkiyenin Etkinlik Turizmi Potansiyelinin Değerlendirilmesi. *Çatalhöyük Uluslararası Turizm ve Sosyal Araştırmalar Dergisi,* Sayı: 2, pp. 35–51.

9 Technology Application in Tourism Events

Case of Africa

Brighton Nyagadza, Tinashe Chuchu and Farai Chigora

Introduction

The coming of the Fourth Industrial Revolution (4IR) worldwide has brought in some new developments commonly known as the "new normal" in tourism events and the entire industry for the African continent. Internet as a common connecting pool of information for many, it has revolutionized the tourism events in Africa and other destinations. Digital and social media platforms have made the African and global visitors to evaluate tourism events, photos, videos and even corporate stories, regardless of where they are located in the world (Hayes et al., 2018). Digitally connected tourism technologies with intelligent systems will revolutionize and optimize digital social media platforms with interconnection of network systems like the Internet of Things (Torrent, 2015; Nyagadza, 2020). Visitors are now able to develop an image about the target tourism events based mainly on the digital and social media platforms, upon their reasoned opinions, emotional interpretation and feelings (Digital Marketing Institute, 2020). This has forced the tourism practitioners in Africa to have a strong drive towards digital and social media platforms leverage in a way to make sure that they can market and sell their tourism events, products and services (Ashley and Tuten, 2015; Jurado et al., 2019).

African tourism events, corporate websites and social media platforms such as Facebook and Google have deep learning algorithms that have been developed with the digital expertise which can have interactivity that allows how tourists travel, see and consume events and services, among other issues (Roces, 2018). As of now Google (and other organizations) has developed robots that are meant to behave like humans and can watch videos on YouTube (Bagot, 2017; Jurkiewicz, 2018). More of the decisions currently being made by human beings (the tourism practitioners) shall be made by the digital algorithms which are much sharper in accuracy, provided there is no unbiased data (Roces, 2018), which might be erroneous (Manyika et al., 2013; Kim et al., 2015). Concerns are more on the ethical side of using the Artificial Intelligence (AI) in tourism events and digital and social media marketing, which has been viewed by scholars such as Cellan-Jones (2014), Garling (2014), Simonite (2017) and

DOI: 10.4324/9781003271147-12

Jurkiewicz (2018), to be dehumanizing the choices of individuals by evidently replacing individual identity with collective, computerized model citizens and employees.

An Overview of Technology Dynamics in Tourism Events

The advent of technologies such as artificial intelligence, virtual reality, robotics, drones and blockchain has had a huge impact on African socio-economic tourism events and activities. The 4IR has been progressing and spreading like wildfire throughout all parts of the African continent, presenting a plethora of opportunities for growth of tourism events (Asghar et al., 2020). However, the onus is on the African developing economies themselves to take advantage of the opportunities presented in order to realize any meaningful tourism benefits from the 4IR boom. If the world's developing economies such as those dominantly found in Africa are to use the advantages presented by the 4IR, they would be compelled to establish proper tourism event structures to address issues to do with adoption, accessibility, affordability and the application of technologies (Adhikari, 2019). One of the fundamental matters that least developed economies need to look into is the early adoption of technologies in tourism events. Leapfrogging and harnessing new technologies as they are introduced can go a long way in positioning Africa tourism events for growth through technology (Lee, 2021). Whilst early adoption of technologies usually creates positive results, slow-moving African states often face the challenge of always following the leader meaning they will always be playing catch-up, therefore remaining one or more steps behind in tourism events development (Lee, 2021). Therefore, it is imperative upon African tourism destinations to identify and take up new technologies to drive the growth of their value chains and ultimately their tourism events.

Africa as a continent still faces the challenge of acquiring technologies for tourism events largely due to barriers such as licence and patent rights. However, with so many open source technologies around, African tourism destinations can circumvent the challenges of patents and licence requirements by harnessing relevant open source technologies (Adhikari, 2019). Open source technologies such as the blockchain can be adopted and adapted to suit particular functional requirements for tourism events (Asghar et al., 2020). The entire tourism events supply chain can be traced using blockchain technology, from the organization to the tourism audience (Black, 2019). A capable human resource is an important factor for African tourism destinations in responding to the 4IR (Lee, 2021). An African tourism destination without enough scientific and technical capacity to conduct tourism events research and provide technical direction cannot realize much out of the technological shift (Kabonga et al., 2021a); and investment in technological infrastructure would count for nothing (Asghar et al., 2020). Therefore, it becomes critical to build capacity and prepare a skilled and knowledgeable human resource to respond to the 4IR demands for the tourism events.

Technology Adoption in African Tourism Events

Information technologies have revolutionized the tourism industry (Buhalis, 2000). This can be said about the Southern African Development Community (SADC), which will form one of the major talking points of this chapter. The SADC is a collective of 16 Southern African states, namely Angola, Botswana, Comoros, Democratic Republic of Congo, Eswatini, Lesotho, Madagascar, Malawi, Mauritius, Mozambique, Namibia, Seychelles, South Africa, United Republic of Tanzania, Zambia and Zimbabwe (United Nations, 2021). The SADC region has a tourism programme which serves as a roadmap to direct and coordinate the development of a sustainable tourism industry (Southern African Development Community, 2020). The SADC tourism program focuses on stimulating visitor movement within the region, maintaining the regions' tourism reputation, improving visitor experience and maximizing partnerships (Southern African Development Community, 2020). Technology has been adopted to some extent in the states mentioned above, however, it is imperative to acknowledge that most of this technology, especially the Internet, has mainly been used in South Africa. It is also important to note that the majority of studies on tourism adoption of technology in the SADC region were quantitative studies with a few taking on a qualitative approach. In Namibia, tourism has played a key role in boosting information technology communications (ICT) and financial services within the county (Wiig, 2003).

The application of technology in tourism within the SADC region has received much interest over the years (Chiutsi et al., 2011; Hottola, 2009; Mpofu and Watkins-Mathys, 2011; Mkwizu, 2019). This adoption of technology in tourism has been explored from various perspectives. For instance, the diffusion and impact of information and communication technology on tourism in the Western Cape, South Africa was investigated by Anwar, Carmody, Surborg and Corcoran (2014). They established that while these new technologies are actively adopted in marketing and booking, more specifically foreign-owned websites have established a dominant command-and-control function, thereby replicating previous patterns of economic success. Wynne et al. (2001) reviewed the impact of the Internet on the distribution value chain of South Africa's tourism, and stressed that the Internet presented a multitude of commercial opportunities for various industries which led them positing the Internet's potential for the tourism industry.

Technology adoption in Zimbabwe was explored by numerous scholars, including Basera and Nyahunzwi (2019), Govere (2013), Madzikatire et al. (2019) and Mupfiga (2015). Some research (Basera and Nyahunzwi, 2019), even went as far as to compare online marketing strategies between South African Tourism (SAT) and Zimbabwe Tourism Authority (ZTA). The ZTA is Zimbabwe's official tourism marketing agency for the country's government (Tripadvisor, 2021), while SAT is the South African government official marketing agency (National Government of South Africa, 2021).

As far as tourism within Zimbabwe is concerned, a study conducted by Mupfiga (2015) focused on ICTs in Zimbabwe's hospitality section. Web services have allowed tourists to act as their own travel agents and not only rely on the Web for gathering information but make their own holiday reservations (Mupfiga, 2015). In the SADC region, Madzikatire et al. (2019) focused on the tourism sector in Zimbabwe and their study qualitatively reviewed technology adoption in Zimbabwe with specific attention to spa tourism technology. This study suggested that spa tourism technology was globally profitable but this was not the case in Zimbabwe. Govere (2013) examined the official website for the ZTA in a study. The main objective of this study was to establish the extent to which this website was able to attract tourists. In 2020, the government of Angola's tourism ministry launched a digital start-up prize with the goal of supporting the digital transformation of Angola's tourism sector (All Africa, 2020). In Zambia, Mwango and Phiri (2018) discuss marketing strategies and sustainable tourism as well as how technology could help the nation's tourism objectives. Musabila (2012) examined the determinants of ICT adoption and usage among small and medium enterprises in Tanzania.

Enhancing the Services Supply Chain for African Tourism Events through Technology

Perfect coordination and monitoring should exist in the African states to enable the flawless flow of tourism events destinations services. For these African tourism events corporations to be managed effectively, it is crucial to have information flowing between them and entities involved in tourism events processes. There are various competitive traditional factors that influence operations of African tourism events, namely labour, capital, land, raw materials, climate, natural environment and arable land. The 4IR technologies have drastically changed the importance of these factors in tourism events execution. Without modern ICT devices (Lechman, 2018), especially the Internet of Things (IoT), big data and cloud computing, smooth operations and development of African tourism events would not be possible (Dima and Maassen, 2018).

A crucial role will be played by capital resources, both physical and human (i.e., knowledge and its application) in the 4IR technologies in the delivery of African tourism events. The research and development (R&D) costs involved in designing, manufacturing, testing and implementing these new technologies and innovations are very high. The effort needed for an innovative idea to be realized and then commercialized remains largely in the hands of African tourism events transnational corporations whose capital affords them comparative advantage. This is evident in the fact that the 4IR has actually been launched by tourism events transnational corporations (TNCs) in the African continent and is being further advanced by major corporations in other nations. However, this suggests that the technological gap between developed and developing countries might widen due to the fact that companies in emerging tourism events destinations in Africa may not have adequate start-up resources to enable the 4IR technologies to be used.

Branding the Technology Tourism Events in Africa

The intensification of technology usage competition in tourism events markets in the African continent has resulted in authorities recognizing destination branding as a tool to use in marketing promotion and management of tourism resources (Kim and Lehto, 2013). Traditionally, tourism events brands have been used to identify general products mainly using names and symbols and now the concept is applied to marketing of services (Pike, 2010). Even with minimal practice in other destinations, the advent of African tourism events branding has resulted in many tourism marketers viewing destinations as brands, such that they have applied the concept from generic product branding theories to destinations (Kim and Lehto, 2013). This shows that traditional identity of tourism events products through the use of brands can still apply to destination marketing. Marketing and technology application for tourism events is done through the use of various objects and designs such as terms, symbols, signs that represent their brands or a combination of all with an advantage of differentiating their own products from those of competitors (Kiliç and Adem, 2012; Nyagadza et al., 2020). It is through the concept of tourism events branding that a unique proposition is created which helps to differentiate products and services provided by one business to those of competitors (Im et al., 2012; Kim and Perdue, 2011). Failure to manage the branding process is detrimental, since the process revolves around brand elements mix, brand identity, brand image building and marketing activities (Im et al., 2012). Therefore, tourism events destination branding is indispensable in contemporary competitive markets dominated by producers and suppliers of homogenous tourism products and services.

There are many existing tourism events destinations in the African continent and globally, offering sometimes similar products and services, which call for differentiation as a marketing strategy, to which branding is a key and promising ingredient (Kim and Lehto, 2013). Tourism events and destination branding also helps in destination marketing through promoting identity of a destination in a global market. Africa's marketing success can be measured by its continental brand performance and improvement in employability of the potential youths (Kabonga et al., 2021b). The current situation shows that the African continent as a destination brand is struggling to establish its market dominance reflected by the continuous re-branding exercises. According to Morrison (2013), a good tourism events and destination brand should be market-tested and well accepted by all stakeholders. The fact that Africa as a destination brand keeps on changing can be due to its failure to positively perform on the global market and not be accepted by pertinent stakeholders.

This is the highest level in tourism events destination brand equity building, whereby tourists develop an attachment and lasting relationship with a destination brand. According to Im et al. (2012), brand loyalty comprises both the attitudinal and behavioural, with the former concentrating on consumers' repurchase intentions and the latter emphasizing the repeat purchasing of a tourism service brand. In tourism events and destination loyalty is seen from repeat visits of tourists, which is triggered by past travel experiences and attachment to tradition (Gartner and Ruzzier, 2011; Kabonga et al., 2021a). Tourists' destination brand loyalty therefore can be assessed according to two main perspectives, namely attitudinal and behavioural loyalty. Behavioural loyalty refers to how consumers behave in their consumption (Mechinda et al., 2010). This is behaviour in purchasing and consumption (Kiliç and Adem, 2012). Attitudinal loyalty is more of emotional attachment that is placed on goods or service (McKercher et al., 2012). It is the composite loyalty (both behavioural and attitudinal) that improves consumption of the tourism products and services, whereby revisiting and recommending others to visit a destination constitutes the complete act (Kiliç and Adem, 2012). When there is frequent repeat visitation and positive word of mouth recommendations, brand loyalty is established (Pike, 2010). The situation in Africa as a destination shows that there is negative tourist loyalty to some of the destination brands, due to tainted images and COVID-19. Tourists' inflows have reduced over the years, which constitutes a sign that that both behavioural and attitudinal loyalty in consumption has become negative.

The Ugly Side of Technology Adoption in Tourism Events in Africa

In addition to this, African tourism events on digital social media marketing sites have been subjectively viewed to be specifically focusing on certain demographics. This has triggered hate groups which might have led to accumulation of information on users and some segments for discrimination, not for tourism events business in Africa (Jurkiewicz, 2018). Further to this, digital platforms such as spam emails and unsolicited ads, encourages prejudice, where the viewers are lured to show hate to some groups through their derogatory speech (Byrne, 2017; Collins et al., 2017; Chiel, 2018). Thus, in the end, the African tourists tend to be limited in turn on how they are likely to use the social media platforms to access information. Ethical challenges of the 4IR emerging technologies in the delivery of the tourism events marketing include, but are not limited to, sufficiency issues related to maturity befitting technological disruption (Van Der Heide and Lim, 2016); and whether there can be affordability to the costs faced for interoperability reasons.

Conclusion

Tourism is a cornerstone of the African continent and particularly SADC economy, together with agriculture and mining with the main contributors being South Africa, Zimbabwe, Botswana, Mozambique, Mauritius, Namibia and Tanzania responsible for 8.8 million jobs (Southern African Development Community, 2020). Digital tourism through mobile phones and the Internet has grown in popularity across Africa (Adeola and Evans, 2019). In Africa the Internet has been actively adopted for enhancing higher education in tourism and hospitality (Tassiopoulos, 2010). This therefore suggests that technology in African tourism is being applied to both practice and academia. The coming in of artificial intelligence, machine learning, availability of big data and algorithms has made it easier to better target the tourism events audience. The post-2000 period in Africa has witnessed great growth in the numbers of social media followers who are very active almost on a daily basis mainly for tourism events. This has lured tourism business operators to shift their marketing efforts to social media marketing, as better leverage to engage tourists for their events together unifying social communities with the commerce and industry fraternity.

Implications to technology application in African tourism events practice

Tourism events capacity building, partnerships to accelerate the response to technological advancement play a key role to overcome African country-level challenges. Developing countries in African need to build and nurture strategic partnerships with enabling partners in tourism key technological sector to improve the implementation of technologies whilst maximizing the relevant available expertise to build capacity for a smooth transformation towards the 4IR dynamics. It is through these partnerships that various African tourism national development agencies can accelerate and harness the potential of digitalization and the 4IR (Yong, 2020). Furthermore, the major concern on ethical tourism events digital and social media marketing are due to the problems related to what is good and ethical in philosophical meanings. Decision making is a key component in any African tourism events organization and there is need to follow a structured way on what is required (Nyagadza et al., 2020). The purpose is not express the ways in which decisions are made, but how to understand the process of ethical decision making in an African tourism events organizational environment. Management responsible for the tourism event's digital and social media marketing are to make specific decision demands knowledge of the tourism in Africa, an assessment of risk and the experience to know the effects to the concerned stakeholders.

References

Adeola, O., and Evans, O. (2019). Digital tourism: Mobile phones, Internet and tourism in Africa. *Tourism Recreation Research*, 44(2), 190–202.

Adhikari, R. (2019). 6 ways least developed countries can participate in the 4IR. *World Economic Forum (WEF)*. Retrieved from: www.weforum.org/agenda/2019/08/6-ways-least-developed-countries-can-participate-in-the-4ir/ (accessed 1 November 2021).

All Africa (2020). *Angola: tourism ministry launches digital startups prize*. Retrieved from: https://allafrica.com/stories/202001290781.html (accessed 20 November 2021).

Anwar, M.A., Carmody, P., Surborg, B., and Corcoran, A. (2014). The diffusion and impacts of information and communication technology on tourism in the Western Cape, South Africa. *Urban Forum*, 25(4), 531–545.

Asghar, S., Rextina, G., Ahmed, T., and Tamimy, M.I. (2020). The Fourth Industrial Revolution in the developing nations: Challenges and road map. *South Centre*. Retrieved from: www.econstor.eu/handle/10419/232222 (accessed 20 November 2021).

Ashley, C., and Tuten, T. (2015). Creative strategies in social media marketing: An exploratory study of branded social content and consumer engagement. *Psychology and Marketing*, 32, 15–27.

Bagot, M. (2017). Google is training robots to understand humans by making them binge-watch YouTube videos. *The Mirror*. Retrieved from: www.mirror.co.uk/tech/google-training-robots-understand-humans-11413196 (accessed 21 November 2021).

Basera, V., and Nyahunzwi, D.K. (2019). The online marketing strategies of the Zimbabwe Tourism Authority (ZTA) and South Africa Tourism (SAT): A comparative study. *Journal of Tourism Hospitality*, 8(3/407), 1–10.

Black, M. (2019). How can developing countries take advantage of the fourth industrial revolution? *Geospatial World*. Retrieved from: www.geospatialworld.net/blogs/how-can-developing-countries-take-advantage-of-the-fourth-industrial-revolution/ (accessed 21 November 2021).

Buhalis, D. (2000). Tourism and information technologies: Past, present and future. *Tourism Recreation Research*, 25(1), 41–58.

Byrne, B.P. (2017). Twitter says it fixed "bug" that let marketers target people who use the N-word. *The Daily Beast*. Retrieved from: www.thedailybeast.com/twitter-lets-you-target-millions-of-users-who-may-likethe-n-word (accessed 10 November 2021).

Cellan-Jones, R. (2014). Stephen Hawking warns artificial intelligence could end mankind. *BBC News*. Retrieved from: www.bbc.com/news/technology-30290540 (accessed 13 November 2021).

Chiel, E. (2018). The injustice of algorithms. *New Republic*. Retrieved from: https://newrepublic.com/article/146710/injustice-algorithms (accessed 18 November 2021).

Chiutsi, S., Mukoroverwa, M., Karigambe, P., and Mudzengi, B.K. (2011). The theory and practice of ecotourism in Southern Africa. *Journal of Hospitality Management and Tourism*, 2(2), 14–21.

Collins, B., Poulsen, K., and Ackerman, S. (2017). Russia used Facebook events to organize anti-immigrant rallies on U.S. soil. *The Daily Beast*. Retrieved from: www.thedailybeast.com/exclusive-russia-used-facebookevents-to-organize-anti-immigrant-rallies-on-us-soil (accessed 17 November 2021).

Digital Marketing Institute (DMI) (2020). *10 trends in digital marketing 2020*. Retrieved from: https://my.digitalmarketinginstitute.com/library/entry/10-trends-in-digital-marketing-in-2020-h1t-a2r (accessed 16 November 2021).

Dima, A.M., and Maassen, M.A. (2018). From waterfall to agile software: Development models in the IT sector, 2006 to 2018. Impacts on company management. *Journal of International Studies*, 11(2), 315–326.

Garling, C. (2014). As artificial intelligence grows, so do ethical concern. *SFGate*. Retrieved from: www.sfgate.com/technology/article/As-artificial-intelligence-grows-so-do-ethical-5194466.php (accessed 10 November 2021).

Gartner, W.C., and Ruzzier, M.K. (2011). Tourism destination brand equity dimensions: Renewal versus repeat market. *Journal of Travel Research*, 50(5), 471–481.

Govere, W.D. (2013). The use of the internet to attract tourists to Zimbabwe: An analysis of the Zimbabwe Tourism Authority website. *International. Journal of Tourism Management and Business Studies*, 3(1), 132–136.

Hayes, J.L., Yan S., and King K.W. (2018). The interconnected role of strength of brand and interpersonal relationships and user comment valence on brand video sharing behaviour. *International Journal of Advertising*, 37(1), 142–64.

Hottola, P. (2009) (ed.). *Tourism Strategies and Local Responses in Southern Africa*. Wallingford: CABI.

Im, H.H., Kim, S.S., Elliot, S., and Han, H. (2012). Conceptualising destination brand equity dimensions from a consumer-based brand equity perspective. *Journal of Travel and Tourism Marketing*, 29, 385–403.

Jurado. E.B., Uclés. F.B., Moral. A.M., and Viruel, M.J.M. (2019). Agri-food companies in the social media: A comparison of organic and non-organic firms. *Economic Research-Ekonomska Istraživanja*, 32(1), 321–334.

Jurkiewicz, C.L. (2018). *Big Data, Big Concerns: Ethics in the Digital Age*. Cambridge: Cambridge University Press.

Kabonga, I., Zvokuomba, K., and Nyagadza, B. (2021a). The challenges faced by young entrepreneurs in informal trading in Bindura Zimbabwe. *Journal of Asian and African Studies (JAAS)*, 56(8), 1780–1794.

Kabonga, I., Zvokuomba, K., Nyagadza, B., and Dube, E. (2021b). Swimming against the tide: young informal traders' survival strategies in a competitive environment in Zimbabwe. *Youth and Society*, https://doi.org/10.1177/0044118X211044524

Kiliç, B., and Adem, S. (2012). Destination personality, self-congruity and loyalty. *Journal of Hospitality Management and Tourism*, 3(5), 95–105.

Kim, D., and Perdue, R.R. (2011). The influence of image on destination attractiveness. *Journal of Travel and Tourism Marketing*, 28, 225–239.

Kim, D., Spiller, L., and Hettche, M. (2015). Analyzing media types and content orientations in Facebook for global brands. *Journal of Research in Interactive Marketing*, 9, 4–30.

Kim, S., and Lehto, X.Y. (2013). Projected and perceived destination brand personalities: The case of South Korea. *Journal of Travel Research*, 52(1), 117–130.

Lechman, E. (2018). *Technological Substitution in Asia*. Abingdon: Routledge.

Lee, K. (2021). *How developing countries can take advantage of the Fourth Industrial Revolution. Industrial Analytics Platform*. Retrieved from: https://iap.unido.org/articles/how-developing-countries-can-take-advantage-fourth-industrial-revolution (accessed 8 November 2021).

Madzikatire, E., Mamimine, P.W., Javangwe, G., and Kazembe, C. (2019). Individuals' experiences with Spa technology consumption in Zimbabwe. *International Journal of Spa and Wellness*, 2(1), 35–52.

Manyika, J., Chui, M., Bughin, J., Dobbs, R., Bisson, P., and Marrs, A. (2013). Disruptive technologies: Advances that will transform life, business, and the global economy. *McKinsey Global Institute*, 180, 17–21.

McKercher, B., Denizci-Guillet, B., and Ng, E. (2012). Rethinking loyalty. *Annals of Tourism Research*, 39(2), 708–734.

Mechinda, P., Serirat, S., Anuwichanont, J., and Gulid, N. (2010). An examination of tourists' loyalty towards medical tourism in Pattaya, Thailand. *International Business and Economics Research Journal*, 9(1), 55–70.

Mkwizu, K.H. (2019). Digital marketing and tourism: Opportunities for Africa. *International Hospitality Review*, 34(2), 5–12.

Morrison, A. (2013). Destination positioning and branding: Still on the slow boat to China. *Tourism Tribune*, 28(2), 1–9.

Mpofu, K.C., and Watkins-Mathys, L. (2011). Understanding ICT adoption in the small firm sector in Southern Africa. *Journal of Systems and Information Technology*, 13(2), 179–199.

Mupfiga, P.S. (2015). Adoption of ICT in the tourism and hospitality sector in Zimbabwe. *The International Journal of Engineering and Science*, 4(12), 72–78.

Musabila, A.K. (2012). *The Determinants of ICT Adoption and Usage among SMEs: The Case of the Tourism Sector in Tanzania*. Amsterdam: Vrije Universiteit.

Mwango, J.K., and Phiri, W. (2018). Marketing strategies and sustainable tourism: An assessment of the Zambia tourism agency. *Marketing*, 4(5), 146–155.

National Government of South Africa (2021). *South African tourism.* Retrieved from: https://nationalgovernment.co.za/units/view/183/south-african-tourism (accessed 20 November 2021).

Nyagadza, B. (2020). Search engine marketing and social media marketing predictive trends. *Journal of Digital Media and Policy.* doi: https://doi. org/10.1386/jdmp_00027_1

Nyagadza, B., Kadembo, E.M., and Makasi, A. (2020). Exploring internal stakeholders' emotional attachment and corporate brand perceptions through corporate storytelling for branding. *Cogent Business and Management*, 7(1), 1–22.

Pike, S. (2010). Destination branding case study: Tracking brand equity for an emerging destination between 2003 and 2007. *Journal of Hospitality and Tourism Research*, 34(1), 124–139.

Roces, M. (2018). Pros and cons of turning all your blogs into vlogs. *INSC Digital Magazine.* Retrieved from: https://theinscribermag.com/pros-and-cons-of-turning-all-your-blogs-into-vlogs/ (accessed 5 November 2021).

Simonite, T. (2017). Two giants of AI team up to head off the robot apocalypse. *Wired.* Retrieved from: www.wired.com/story/two-giants-of-ai-team-up-to-head-off-the-robot-apocalypse (accessed 4 November 2021).

Southern African Development Community (SADC) (2020). *Tourism programme 2020–2030.* Retrieved from: www.sadc.int/files/9715/8818/8701/SADC_Tousim_Programme_English.pdf (accessed 21 November 2021).

Tassiopoulos, D. (2010). *Use of the internet for enhancing tourism and hospitality higher education in Southern Africa: Implications for e-learning.* Retrieved from: www.intechopen.com/chapters/10548 (accessed 4 November 2021).

Torrent, J. (2015). Knowledge products and network externalities: Implications for the business strategy. *Journal of the Knowledge Economy*, 6(1), 138–156.

Tripadvisor (2021). *Zimbabwe Tourism Authority.* Retrieved from: www.tripadvisor.co.za/Attraction_Review-g293760-d13155642-Reviews-Zimbabwe_Tourism_Authority-Harare_Harare_Province.html (accessed 20 November 2021).

United Nations (UN) (2021). *National implementation of agenda 21.* Retrieved from: www.un.org/esa/earthsummit/sadc-cp.htm (accessed 20 November 2021).

Van Der Heide, B., and Lim, Y-S. (2016). On the conditional cueing of credibility heuristics: The case of online influence. *Communication Research*, 43(5), 672–93.

Wiig, A. (2003). Developing countries and the tourist industry the Internet age: The case of Namibia. *Forum for Development Studies*, 30(1), 59–87.

Wynne, C., Berthon, P., Pitt, L., Ewing, M., and Napoli, J. (2001). The impact of the Internet on the distribution value chain: The case of the South African tourism industry. *International Marketing Review*, 18(4), 420–431.

Yong, LI. (2020). Making the Fourth Industrial Revolution work for all. *UNCTAD.* Retrieved from: https://unctad.org/news/making-fourth-industrial-revolution-work-all (accessed 10 November 2021).

Part III

Technology Application in Tourism Events

Case of Australia

10 The Ubiquitous Role of Mobile Technology Application in the Australian Open

Hasanuzzaman Tushar, Shafiqur Rahman, Sweta Thakur and Md Sazzad Hossain

Introduction

The Commonwealth of Australia is a continent country with a nearly 26 million population and is considered as the world's 13th largest economy (World Bank, 2020). Tourism is a vital part of the Australian economy which consists of both international and domestic components. The attraction of Australian tourism has made it the 13th largest tourism market in the world in 2019, contributing to a 3.1% share of global tourism incomes, which is currently worth US$1.3 billion (Austrade, 2020). During 2019, Australians spent $80.7 billion and $26.3 billion on domestic overnight trips and day trips respectively.

The country has vibrant cities, magnificent natural landscapes, coastlines with surfing opportunities, native wildlife including kangaroos, koalas, dolphins, whales and rainbow loricates, and friendly multicultural citizens make Australia one of the most attractive tourist destinations in the world. Every year, destinations like Sydney, Brisbane and Melbourne attract a large number of tourists. Also, other major tourist attractions include the Gold Coast, Queensland and the Great Barrier Reef. According to Austrade (2019), tourists in Australia grew to 9.4 million during 2018–19 from 5.5 million during 2008–09. According to the Australian Bureau of Statistics, there were 611,700 tourism jobs in Australia at the end of June 2020 (Austrade, 2021).

Sport is a vital part of Australian culture and plays a significant role in drawing a large number of overseas tourists every year (Bateman et al., 2020). Sports lovers fly to Australia to enjoy the incredible sports events, starting from the Australia Open to major marathons. Other major events in Australia that attract tourists include AFL Grand Final, Australian Grand Prix, Melbourne Cup, Rolex Sydney Hobart Yacht Race, World's Women Cups and Canberra Flower Show. Overseas tourists attending the above events also contribute to the various industries including aviation, hotel/motel, restaurant, grand transportation and accommodation industries (Pandey, 2020).

DOI: 10.4324/9781003271147-14

The Australian Open

The Australian Open is an annual tennis tournament that takes place in the last two weeks of January in Melbourne (Sánchez-Pay et al., 2021). This event is the first of the four annual Grand Slam tennis championships, before the French Open, Wimbledon and the US Open. Its features include men's and women's singles; men's, women's and mixed doubles; furthermore, this event shows junior's championships, wheelchair, legends and exhibition events. The Australasian Championship, currently known as the Australian Open, first held in 1905, has become one of the leading sporting events in the Southern Hemisphere (Quayle et al., 2019). Then, in 1969, it became the Australian Open. In 2020, more than 812,000 people attended the Australian Open (BBC Sport, 2021). Tennis Australia manages the Australian Open and organizes the event at the Albert Reserve Tennis Centre (Tennis Australia, 2021). Tennis Australia also promotes tennis and its participation; facilitates player development; stages local and international events and invests in tennis facilities throughout Australia.

The Role of Mobile Technology on Australian Open

The recent unprecedented growth of technology has altered the fundamental ways of the tourism industry and the experiences of tourists. The wave of these technological changes is the result of the evolution of mobile technologies and their wider ranges of applications. On one hand, service providers are experiencing the benefits of having visibility in different mobile-based applications to provide services and prompt communication with the target consumers. On the other hand, mobile-based technology offered a variety of choices to the tourists to evaluate and make the preferences at the pre-trip and during trip stages. Mobile technology has also enabled payment, real-time assistance and feedback for both consumers and service providers (Hossain et al., 2021). This chapter has focused on the mobile technology role of a particular event "Australian Open" which is explored under different viewpoints including consumers or tourists and service providers.

Australian Open is attracting an increasing number of tourists in Australia who prefer to use various types of mobile applications to share their experience in social media, book their accommodations/transportations online and make their payments using e-commerce platforms, order online food delivery and so on. The Australian Open also has its presence in all popular social media platforms (i.e., Facebook, Twitter, Instagram, and a few others). Digital technologies including various mobile apps have the potential to increase the efficiencies of tourism for the travellers who attend the Australian Open (Dhakal, 2021). A few such technologies for the travellers/service providers of Australian Open are discussed below.

Australian Open tickets via mobile

Nowadays, at the Australian Open (AO), mobile tickets have replaced traditional paper tickets (Ticketmaster, 2021). Mobile tickets contain seat details, barcodes and entry details. In addition, mobile tickets also can contain other pertinent and significant information. These tickets are dynamic which enables them to update whenever required. Most notably, mobile tickets operate within COVID-19 safe guidelines. All attendees are required to hold their tickets on their mobile devices. All AO 2021 patrons attending the event are required to assign each ticket to a ticketholder using the Ticket Forwarding function.

Live streaming via mobile phone

The current COVID-19 situation provided various levels of restrictions to maintain social distance. One of Australia's most popular television channels, Channel 7, live streams the Australian Open via a dedicated mobile app (Ausopen.com, 2021). The Livestream service is available via any Internet-enabled device, including mobile phones. This unique service allows the audience to watch the sport on their mobile phones wherever they are. The Livestream can be accessed by visiting the website "7Plus" or running the 7Plus app on a mobile device (Australian Open Championships, 2021). However, the free Livestream is geo-blocked beyond Australian territories, so the access to Channel 7 will be blocked once someone departs the country.

Hi-tech "selfies" at the tennis

While the visitors are at the Australian Open tournament venue, they can capture pictures/images/selfies with smartphones. Fans can also take various still pictures as well as videos of various actions during the tournament and immediately share them with family and friends by sharing via social media. Examples are the images of coaches shaking hands before the game, or the players on court, at the foul line, the service line, or shaking hands after the game. The audience enjoys their mobile phones by capturing historical images for their records or to share with family and friends. Furthermore, at the Rod Laver Arena, the main venue of the Australian Open, "Fan Cam" enables visitors to use the cameras of the venue to snap selfies. Fans can visit a mobile site using their smartphone, enter their seat number and ask one of the arena's cameras to take a photograph, which will then be sent to their smartphones (Lebel and Danylchuk, 2019).

Use of accommodation booking app

Accommodations are available near the venues of the Australian Open including Crown Promenade Melbourne, Quay West Suites Melbourne, Hotel Lindrum Melbourne, Sofitel Melbourne on Collins and Sheraton Melbourne Hotel. Booking.com. Travellers attending the Australian Open may consider using Expedia.com.au, Agoda.com, Hotels.com, Accor Hotels and so on, accommodation apps accessing from their mobile phones. Also, various apps are available to explore local tourist attractions, meals and other entertainments nearby (Krause and Aschwanden, 2020).

Ridesharing apps in Australia

Domestic and overseas tourists who attend the Australian Open book ridesharing vehicles using their mobile apps (Bishop, 2019). To book rideshare, one needs to download the app of choice, open it up and select the destination. Passengers are alerted when a driver accepts the request, the driver is on their way, or arrive at the pickup location. On Melbourne roads, the most well-known ridesharing taxi is Uber that Australian Open travellers can use between airport and hotel or between any other destinations. In addition to Uber, other ride-sharing app-based taxi companies are Didi, Ola, GoCatch and Shebah. During the Australian Open tournament, thousands of visitors use the local ride-sharing apps and services.

Restaurants/catering

Australian Open fans dine at the local restaurants and also make orders to the food delivery services like Uber Eats (Poelman et al., 2020). These travellers use mobile apps to book restaurant tables as well as use the apps to order food from the restaurants while eating inside their hotel rooms. Such mobile apps are very useful and convenient to book restaurant tables and to order food to deliver at a certain destination. During the Australian Open tournament, a large number of tourists use these apps and their services, impacting the local economy.

Cashless payment

The use of cashless payment apps like e-wallet and iPay has become popular in recent times (Chinnasamy et al, 2021). Especially during the COVID-19 pandemic period, everyone is trying their best to make a cashless payment. The possibility of virus transmission via cash has encouraged people around the world to make cashless payments, including in Australia. Almost all hotels, restaurants and other service providers are now receiving cashless payment via contactless terminals. Australia is moving towards becoming a cashless country, which tourists will enjoy during the new normal.

Perspectives of Service Provider

For any business, prolonged success is determined by well-pleased customers. The tourism organizations are nothing unusual where the complaints from the customers can have a remarkable impact on the success of the business. There are various reasons which includes dissatisfaction with the service they have received, ill-mannered behaviour from the service provider, too much waiting time and unmet expectations. However, it is difficult, if not impossible, to satisfy customers all the time. Hence, it is very important to recognize complaints and apply some appropriate retrieval tactics and undertake to do something about the problem. The reason for service failure and the recovery from the service provider's perspective by subsequent follow-up to these complaints is helpful to overcome the issue. The focus is to investigate what relevant measures the service providers themselves thought of and can be taken to address customer complaints. The other issue is that the customer engages with the service provider, and also because they are interacting with several service providers. Many services cover tourism and travel-related services which includes services offered by the tour operator, travel agencies, tourist guides, hotels, motels, restaurants and many other related services.

Strategic and operational management

The functioning of strategic management plays a vital role in tourism. The definition of strategic planning and the existence of a development strategy is very challenging and cannot be denied. Strategic management is an enterprising process of achieving a long-term goal to check the similarity of the corresponding field in the intended tourism atmosphere. The operations management in the tourism industry owes much to systems thinking and helps to analyse and improve the decisions to be made in an organization. It includes the service delivery quality standards, performance standards, capacity, control and so on.

The implementation of priority setting goals for tourism is to be set by the national economy. The resource management, management information systems, enlarging of the tourism business, overseeing the novel and innovative tourism projects help the strategic management to maintain the efficiency and productiveness of the long run progress growth in tourism. Therefore, to avoid the threat of the development of tourism it is necessary to have successful parts of operations in form of strategic and operational management. It has been identified and emphasized that for long-term orientation the strategic decisions taken exclusively at the highest level should always relate to different organizational levels for much more effective results as it can affect them in various ways. With a huge demand from foreign investors, the need for operational managers will be even more noticeable. In addition to the substantial experience in the pertinent field with managerial aspects and the knowledge of the local population, geography, demand, competition and so on, there may be a high risk for the foreign investor. Only capable operational management can conduct the appropriate objectives set in the development of the tourism sector (Kirovska, 2011).

Revenue management

Richard and William (2015) stated that over the past 50 years, revenue management has played a remarkable role in the progress of many industries, including tourism by enhancing the theory and applications in airline, hotel, cruise ship and other tourism sectors. The most common factors involved are rational decisions taken for real-time booking, key elements as customer behaviour, demand foreseeing, revenue and cost factors, functional reflection and so on, which helps to maximize the revenue and hence profit. The management of revenue management is an essential concept in the hospitality industry, as the service providers can predict demand and enhance the relevant factors like availability and pricing to accomplish the optimum financial results. The future of revenue management in tourism will get affected by the strategy the organization is following, excessive levels of competition, technological impacts and logical capabilities (Revfine.com, 2021). "Selling the right room, to the right client, at the right moment, for the right price, through the right distribution channel, with the best cost efficiency." At first, the airline industry introduced the concept of revenue management and the idea of dynamic pricing came up. However, significantly it has been identified as applicable in any industry where the anticipation can be done for different customers willing to pay different prices for similar products. For constructive revenue management, the business must always be ready for any adjustments to be made based on customer spending habits and speculating the demand. The strategy should follow the most common factors and the analyses for these factors are very critical such as bookings data, pre and existing weather forecasts and competitor data.

Online marketing and promotion

Basic marketing strategies are also applicable for tourism marketing and are designed to attract visitors to a definite location including hotels, motels, cities, states, consumer attractions and convention centres. The demand for these venues is correlated with consumer demands. To compete, marketing should include determining the unique selling benefit or other convenience over its competitor in the tourism industry. They might think of offering people who are seeking business and pleasure together, travel comfort, event venues, interesting activities and nightlife from the adults and children point of view. Hence, tourism marketing uses diverse ways of communications schemes and methodologies to promote areas and destinations. There are various ways to do these promotions which may incorporate advertisements in trade magazines, dealing and cooperating with the event management industries, building websites and placing ads in consumer publications. Also, promotional offerings include brochure packets, discount coupons and other relevant materials to the potential visitors (Sofronov, 2019).

The change has been noticed in the costumer behaviour because of the promotion of the Internet. As per the classic theory of AIDA (attention, interest, desire and action) in tourism marketing, it is obvious that the tourists have used digital (or Internet) technologies at all levels, although of various rules. The calculations of AIDA are made at different stages, which are: 77% seek information from the Internet; 65% desire; 34% action. The growth of Internet booking is remarkable: it increases to 244% during 2012–2015. According to the current statistics, over 90% of travellers do their research online, and finally, 82% of them make their booking online as well (Pitana and Pitanatri, 2016). For the long run considering the trend and the future of digital marketing, all the relevant services are now using digital marketing which includes countries, tourism destinations and travel agencies; however offline marketing is still endorsed.

However, because of the remarkable benefits and characteristics of digital marketing which cannot be easily found in offline marketing, the growth of Online Travel Agents (OTA) is extremely high, both in terms of number and volume. There is no hurdle to reach the marketers using digital marketing, one can easily reach the global audience, personal communication up to individual level can be improved, can deliberately reach certain targets (Pitana and Pitanatri, 2016).

However, the use of marketing skills and innovative techniques is very important to understand the market needs, understanding the behavioural change because of communication. It has been noticed that designing more sustainable products and use of the same in the travel industry can effectively increase the business (Font and McCabe, 2017).

In addition, it has also been identified that technologies such as artificial intelligence (AI) or virtual reality (VR) may play an increasingly important role in the medium term. The online information sources will surpass conventional travel agencies, except for specialized and advisory services (Toubes et al., 2021).

Guest service

To achieve success in tourism is to attract more costumers and to take care of them and impress them as the industry is dominated by customer service. For that, it is necessary to understand the customer needs and how to successfully meet those needs. The complexity grows with the demands of the customer and delivering the experience grows difficult as well. Various factors need to be considered such as the development of guest service, the major areas like food, lodging, events and so on. It also involves the implementation of strategic planning, staff development, and marketing as quality service is evident in all the operations, its people and its plan. Though it is a bit challenging at the same time it is a great opportunity for the service provider to supply as per the demands for gaining a lifelong experience. The information, collection and data are very useful and help in an effective business (SiteMinder, 2021). Guest service is like from pre-booking to post-stay which involves various scope to meet the customer needs, which not only helps to keep the old guests as well as to attract the new guests by providing all possible relevant services where the focus is on quality, reliability, sensitivity, visibility and guarantee (Bagdan, 2013).

E-Security and apps design

It is important to understand and explore the patterns of consumption of ICT and its application in the tourism and hospitality industry (THI) which might not be a new occurrence in advanced countries but is of practical use, especially in developing countries. The factors associated with ICT include e-marketing, e-strategic management, ensuring security and rendering tourism and hospitality services that are relevant to any country. While talking about the application, the finding is that it should acknowledge five factors: Organizational, Technical, Economic, Environmental and Personal, and how the outcome can enhance the current problems ICT applications can overcome by developing new policies and procedures in these industries (Sardar et al., 2021).

Technology is everywhere in our day-to-day life, it is navigating the world and making our life better each passing day. Linking technology with tourism magnifies and expands its attributes and makes the process hassle-free for the tourists. The use of Quick Response (QR) codes as well as Near Field Communication (NFC) technology are massively used in inventory systems, online payment systems, and are tied with the tourism industry to enhance the business and its effectiveness. These features using a mobile device that can assist the tourists by providing much-needed information about tourist attractions and hence help in making their travelling experience a trouble-free and amusing journey. But it can come up with a risk of malicious substitution and proper strategy is required to address those issues (Ekundayo et al., 2020).

Tourism means not only should we be thinking of marketing only, but also need to consider the people, data, delivery of those data with the help of different platforms, and mobile location-based marketing. Such a marketing approach includes social media marketing, push notifications, e-newsletters and even off-line marketing that helps the business to be more effective and can target the customers easily (Yim et al., 2019). But the success of any application is directly proportionate to the degree of user acceptance. So, before the development process, it is necessary to explore user expectations to make a successful product. User experience design (UXD) is the research arena that identifies the needs of the users, their expectations and acceptance. Mobile flight booking ticket application (MFBTA) has got various types of transactions and worked on the UXD strategy (Yazid and Jantan, 2017).

Sustainability

It is very important to identify the gaps between sustainability-related theories and observations at business organizations and recommends accordingly to overcome those gaps in the tourism industry. The findings include the future research initiatives on green business modelling for sustainability from the economic, environmental and social performance point of view can be the motivation for sustainable tourism (Pushpakumara et al., 2019).

There is a serious threat to the natural environment because of the various forms of environmental problems and this is currently affecting the tourism and hospitality industry. To better understand sustainability in the environment it is important to understand environmental protection, fostering consumer behaviour, which involves association with nature, green products, environmental knowledge and its importance in everyday life, environmental corporate social responsibility and so on, while in turn keeping a track of universal goals for recommending sustainability. There are two key factors based on which sustainability can be measured and those are corporate strategy and the potentiality of marketing, which can impact the corporate decisions. The bottom line is to design and sustainable market products. Also, to understand the consumer behaviour and identify the ways to change this behaviour so that they tend to have sustainable choices without compromising their main incentive. Though there are many challenges, many innovative solutions have been developed to overcome sustainability challenges. There is a growing inclination for the development of additional sustainable tourism products and services that can be advertised successfully while maintaining goals for a robust, successful and long-term outcome, environmentally accountable industry and marketplace in the tourism industry and related destination services discussed in the Journal of Sustainable Tourism (e.g. Han, 2021, p. 879).

Acceptance, Cost and Affordability of M-tourism Application in Australian Open

M-tourism is one of the key terminologies in the tourism industry. Most people use mobile for using the Internet because of smartphone availability. In addition, mobile tourism (M-tourism) applies to mobile devices such as smartphones, tablet computers, smart bands, GPS locator/beepers and related tourist services software. People use mobile Internet to check email, inform about the weather, news, investigation, track travel itineraries and so on.

Mobile devices are usually furnished with 4G and 5G data transmissions, Wi-Fi data communication and GPS features to detect locations. Moreover, 40% of foreign travellers presently have an Internet access smartphone (i.e., iPhone). Smartphones provide a variety of input capabilities and have substantial screen sizes. Smartphones enable dozens of mobile apps (Wang et al., 2012). The travel app sends its contents to the user when the tourist reaches an attraction. The tourist can get more efficient information in good time. The expert opinions extend the assessment of mobile tourism services. The mobile application is still essential and especially apps about travel information, communication and distribution.

Besides, the Technology Acceptance Model (TAM) is a widely used model to justify user acceptance behaviour (Granić and Marangunić, 2019). This present study investigates the mobile application used to travel in the Australian tourism industry. The progress in mobile technology has led to an increasing number of mobile device users. Many studies on the use of mobile devices have likewise rapidly evolved. The new mobile commercial (m-commercial) functions in a much different environment than wired Internet e-commerce because of the specific qualities and limitations of mobile devices and wireless networks. As a communications and marketing channel, mobile communications are the essence of both e-commerce and travel.

Some wireless infrastructures simultaneously enable data transmission to all mobile users in a particular region. This is an effective way of disseminating information to a vast number of tourists. Tourists download a travel app to their smartphone and can obtain location-based travel information and get directions. The travellers' acceptance of the App-Based Mobile Tour Guide (AMTG) is a current issue in the tourism sector. Thus, the uses, capacity or environment in which tourists can get easy access indicate the acceptance.

Moreover, the development of mobile applications and websites, especially for tourism service providers, involves considerable costs which are typically made available to third-party web marketing firms, generating further, cost-effective issues in support, maintenance and up-to-date information. The most effective cost option seems to have been to provide conventional desktops and an additional mobile adapter, or, in some instances, a website responsive to the problem because immersive, emotive and valuable sites are already well developed by tourist service providers. The concept approach is logic overall, and the usability of the process takes precedence with growing task complexity. Form fields loosen data previously input, pop-up windows on smartphones and unknown navigational structures. The bookings are significant usability showstoppers that developers should tackle seriously and prevent at all costs.

Furthermore, tourism organizations may reach tourists wherever at any time using mobile devices. On the other hand, tourists also may access whatever information they want via Internet-enabled mobile devices, any time they want the information wherever they are. In this sense, m-tourism provides a service or application any time a demand arises.

Mobile device owners typically need distinct apps and services so that m-tourism applications may tailor information or services to a specific user. Travellers utilize the Internet more to get comprehensive information about tourism products and services. Buhalis (2019) also projected the Internet to be the leading distribution channel for mobile devices. The mobile device may provide location-based services enabling passengers (a) to identify people and things and tours, (b) to construct routes (i.e., travelling there); (c) to find nearby objects such as restaurants, stores, hotels, and attractions. Due to the inherent portable mobile devices, users may participate in other activities such as meeting others or travelling while making transactions or getting information via their Internet-compatible mobile devices. Thus, these facilities make it affordable for tourists to use mobile technology for travel purposes or tourism destinations. Sometimes, a few tourists cannot access to mobile technology in order to access information. Thus, tourism destination/management should place a guide for helping to find that information.

Conclusion and Implications

The present study has investigated the mobile technology application used in the Australian Open. Thus, this study has identified important theoretical implications that indicate the consequences and specify the existing features. The above literature, discussion and findings have shown that mobile technology in the tourism sector is widely recommended for both host and tourist. Several studies have postulated the mobile technology application and suggested various ways, such as apps and websites (Buhalis, 2019). Based on the suggestions, mobile technology gives destinations new opportunities and difficulties to capture tourist attention and successfully communicate marketing messages to specific audiences. Tourism organizations/destinations are currently attempting to improve their knowledge of the demands and preferences of mobile services.

This study also emphasizes the managerial implication. Based on the above discussion, the destination management/authority should take necessary action in enhancing mobile application uses/acceptance. The destination management also should care about tourist behaviour of using technology or accessing the information. It is noted that the degree of perceived value and perception are key factors affecting mobile travellers' happiness with their mobile experiences. In conclusion, the present study added significant value for the tourism destination to formulate tourist dependency on the service and product delivery. Incorporating mobile technology acceptance/uses application stretched with additional variables provided a model with a theoretical basis to explain behavioural intention/revisitation. This approach will provide a preliminary proposal for further study of TAM (Technology Acceptance Model) models.

References

Ausopen.com (2021). *Watch live.* Retrieved from: https://ausopen.com/watch-live (accessed 15 September 2021).

Austrade (2019). *China takes top spot in Australia's tourist rankings.* Retrieved from: www.austrade.gov.au/news/economic-analysis/china-takes-top-spot-in-australias-tourist-rankings (accessed 12 September 2021).

Austrade (2020). *Australia's "big five": The Dynamic Industries That Power Trade & Investment.* Retrieved from: www.austrade.gov.au/news/economic-analysis/australias-big-five-the-dynamic-industries-that-power-trade-investment (accessed 20 September 2021).

Austrade (2021). *Tourism employment in Australia.* Austrade. Retrieved from: www.austrade.gov.au/australian/tourism/policy-and-strategy/labour-and-skills (accessed 21 September 2021).

Australian Open Championships (2021). Watch 2021 Australian Open Championships Online: Free Streaming & Catch up TV in Australia. *7plus.* Retrieved from: https://7plus.com.au/2021-australian-open-championships (accessed 20 September 2021).

Bagdan, P. (2013). *Guest Service in the Hospitality Industry.* Hoboken, NJ: John Wiley & Sons, Inc.

Bateman, J.E., Lovell, G., Burke, K.J., and Lastella, M. (2020). Coach education and positive youth development as a means of improving Australian sport. *Frontiers in Sports and Active Living,* 2, 180.

BBC Sport (2021). *Australian open to allow up to 30,000 fans to attend per day.* Retrieved from: www.bbc.com/sport/tennis/55866539 (accessed 20 October 2021).

Bishop, R. (2019). The "Sharing Economy" and the Uber Evolution in Australia. *e-Journal of Social & Behavioural Research in Business,* 10(3), 34–40.

Buhalis, D. (2019). Technology in tourism-from information communication technologies to eTourism and smart tourism towards ambient intelligence tourism: a perspective article. *Tourism Review,* 75(1), 267–272.

Chinnasamy, G., Shrivastava, P., and Siju, N. (2021). Diffusion and adoption of e-wallets in Oman for sustainable growth. In N.R. Al Mawali, A.M. Al Lawati and Ananda S. (Eds.), *Fourth Industrial Revolution and Business Dynamics.* Singapore: Palgrave Macmillan, pp. 177–198.

Dhakal, S.P. (2021). Accommodation and food services. In *The Fourth Industrial Revolution* (pp. 73–92). Singapore: Springer.

Ekundayo, S., Baker, O., and Zhou, J. (2020). QR Code and NFC-Based Information System for Southland Tourism Industry-New Zealand. In *2020 IEEE 10ʰ International Conference on System Engineering and Technology (ICSET)*. IEEE, pp. 161–166.

Font, X., and McCabe, S. (2017). Sustainability and marketing in tourism: Its contexts, paradoxes, approaches, challenges and potential. *Journal of Sustainable Tourism*, 25(7), 869–883.

Granić, A., and Marangunić, N. (2019). Technology acceptance model in educational context: A systematic literature review. *British Journal of Educational Technology*, 50(5), 2572–2593.

Han, H. (2021). Consumer behavior and environmental sustainability in tourism and hospitality: A review of theories, concepts, and latest research. *Journal of Sustainable Tourism*, 29(7), 1021–1042.

Hossain, S.F.A., Shan, X., Nurunnabi, M., Tushar, H., Mohsin, A.K.M., and Ahsan, F.T. (2021). Opportunities and challenges of m-learning during the COVID-19 pandemic: A mixed methodology approach. In J. Zhao and J. Richards (Eds.), *E-Collaboration Technologies and Strategies for Competitive Advantage Amid Challenging Times*. Hershey, PA: IGI Global, pp. 210–227.

Kirovska, Z. (2011). Strategic management within the tourism and the world globalization. *UTMS Journal of Economics*, 2(1), 69–76.

Krause, A., and Aschwanden, G. (2020). To Airbnb? Factors impacting short-term leasing preference. *Journal of Real Estate Research*, 42(2), 261–284.

Lebel, K., and Danylchuk, K. (2019). *Tennis and social media*. In R. Lake (Ed.), *Routledge Handbook of Tennis: History, Culture and Politics*. Abingdon: Routledge, pp. 329–337.

Pandey, A., Sahu, R., and Joshi, Y. (2020). Kano Model Application in the Tourism Industry: A Systematic Literature Review. *Journal of Quality Assurance in Hospitality & Tourism*, DOI: 10.1080/1528008X.2020.1839995

Pitana, I.G., and Pitanatri, P.D.S. (2016). Digital marketing in tourism: the more global, the more personal. *International Tourism Conference: Promoting Cultural and Heritage Tourism, Udayana University*, Bali, 1–3.

Poelman, M.P., Thornton, L., and Zenk, S.N. (2020). A cross-sectional comparison of meal delivery options in three international cities. *European Journal of Clinical Nutrition*, 74(10), 1465–1473.

Pushpakumara, W.H., Atan, H., Khatib, A., Azam, S.F., and Tham, J. (2019). Developing a framework for scrutinizing strategic green orientation and organizational performance with relevance to the sustainability of tourism industry. *European Journal of Social Sciences Studies*, 4(3), 1–18.

Quayle, M., Wurm, A., Barnes, H., Barr, T., Beal, E., Fallon, M., et al. (2019). Stereotyping by omission and commission: Creating distinctive gendered spectacles in the televised coverage of the 2015 Australian Open men's and women's tennis singles semi-finals and finals. *International Review for the Sociology of Sport*, 54(1), 3–21.

Revfine.com (2021). *Knowledge platform for the Hospitality & Travel Industry*. Retrieved from: www.revfine.com/ (accessed 12 October 2021).

Richard, B.M., and W.P. Perry (2015). Revenue management. In J. Jafari and H. Xiao (Eds.), *Encyclopedia of Tourism*. Cham: Springer International Publishing, pp. 797–98.

Sánchez-Pay, A., Ortega-Soto, J.A., and Sánchez-Alcaraz, B.J. (2021). Notational analysis in female Grand Slam tennis competitions. *Kinesiology*, 53(1), 154–161.

Sardar, S., Hossain, M.E., Kamruzzaman, M., and Ray, R. (2021). ICT applications in tourism and hospitality industry of Bangladesh: A research review. *IJRDO - Journal of Business Management*, 7(6), 64–74.

SiteMinder (2021). *The world's Largest Open Hotel Commerce Platform.* Retrieved from: www.siteminder.com/ (accessed 10 October 2021).

Sofronov, B. (2019). The development of marketing in tourism industry. *Annals of Spiru Haret University. Economic Series*, 19(1), 117–127.

Tennis Australia (2021). *About Tennis Australia.* Retrieved from: www.tennis.com.au/about-tennis-australia (accessed 20 October 2021).

Ticketmaster (2021). *No pdfs, no paper, no hassle.* Retrieved from: www.ticketmaster.com.au/mobile (accessed 20 October 2021).

Toubes, D.R., Vila, N.A., and Brea, J.A.F. (2021). Changes in consumption patterns and tourist promotion after the COVID-19 pandemic. *Journal of Theoretical and Applied Electronic Commerce Research*, 16(5), 1332–1352.

Wang, D., Park, S., and Fesenmaier, D.R. (2012). The role of smartphones in mediating the touristic experience. *Journal of Travel Research*, 51(4), 371–387.

Yazid, M.A., and Jantan, A.H. (2017). User Experience Design (UXD) of mobile application: An implementation of a case study. *Journal of Telecommunication, Electronic and Computer Engineering (JTEC)*, 9(3–3), 197–200.

Yim, J., Ganesan, S., and Kang, B.H. (2019). Location-Based Mobile Marketing Innovations 2018. *Mobile Information Systems*, DOI: 10.1155/2019/2164708

Part IV
Cases from Europe

11 Technological Innovations in Event Sport Tourism

Case Study of the 2021 Sabre World Cup in Budapest in Hungary

Katalin Csobán

Introduction

The unprecedented growth of technological innovations fundamentally changed the tourism industry in the last decades. Tourism and hospitality businesses strive to take advantage of the new opportunities provided by the development of online digital technologies, such as the Internet, wireless and mobile technologies, audio and video streaming, and so on, while they have to face new challenges arising from the unstable environment and the fast pace of change.

The assimilation of online digital technologies has also altered the way tourism events are prepared, promoted and implemented, which was further accelerated by the present coronavirus pandemic. The interaction with the other participants of an event became feasible with the help of the various social media platforms, which may contribute to an enhanced travel experience as well as a sense of community. Sport events are also greatly influenced by technological innovations, which seems to take sports competitions to a new dimension. The literature review below commences with an overview of the sports tourism and some of the major technological advancements supporting sports events. Then the chapter focuses on a professional sport event – the individual and team Sabre World Cup of 2021 organized in Budapest, Hungary, which incorporates both active participation on-site and passive participation online.

Sports Tourism: A Conceptual Approach

There is a close connection between tourism and sports, as is clearly demonstrated by the large number of travellers who make their journeys to participate actively or passively in competitive or recreational sport activities. Sport may be either the primary or secondary travel motivation or it may serve as an additional recreational activity forming part of the travel experience (Gammon and Robinson, 2003). Definitions of sport tourism suggest that the demand side of the sport tourism market consists of athletes and their teams (coaches, referees, and so on)

DOI: 10.4324/9781003271147-16

as active sport tourists, as well as spectators or fans as passive sport tourists (Hinch and Higham, 2011).

Sport tourism is by definition a "sport-based travel away from the home environment for a limited time where sport is characterized by unique rule sets, competition related to physical prowess and play" (Hinch and Higham, 2001, p. 49). This type of tourism can be classified as active, nostalgia and event sport tourism, which are not sharply distinct categories (Gibson, 2005).

Active sport tourism refers to participatory sports-related travel, which includes mega events (e.g., the Olympic Games, FIFA World Cup or recurring small-scale events such as local friendly tournaments) and non-event sport tourism (e.g., golf, skiing) (Kaplanidou, 2010).

Nostalgia sport tourism covers visits to sport museums, halls of fame, themed bars and restaurants, heritage events and sports reunions, and it is known to bear some similarity to heritage tourism (Gibson, 2005).

Event sport tourism typically involves travel to a sporting event, where the number of spectators frequently exceeds the number of the elite athletes. In other cases the non-elite competitors may outnumber the spectators, who may be the sportsmen's families and friends. Previous research suggests that non-elite participants are more willing to take part in tourist activities, for example in sightseeing or eating out at the destination of the sports event (Hinch and Higham, 2011).

The scale of sport events ranges from small-scale local events, where athletes may outnumber the spectators, to mega-events attracting millions of visitors to the destination. Although small-scale sports events may attract fewer visitors to a destination, they still have important potential for tourism. They often exert a significant positive influence on the local community by providing additional incomes, using the existing infrastructure, reducing seasonality, raising local pride and community spirit. The size of the event and the way in which a sport event is hosted has implications for the sustainability of the event and that of the destination. Sport tourism related to small-scale events as opposed to mega-events is generally considered to be a more sustainable form of tourism (Gibson et al., 2012; Csobán and Serra, 2014).

Sports events may have certain disadvantages, which seem to increase with the scale of the event. Mega sports events often require infrastructure development provided by public funding, moreover the operations and maintenance of the facilities may become unsustainable after the event (Hiller, 2006). Mega sporting events generate a higher influx of people, causing overcrowding, infrastructure congestion, environmental degradation and disturbance to the host population (Chernushenko, 1996).

On the positive side, hosting large-scale international sports events has certain economic benefits resulting from tourists' expenditure associated with the event (Burgan and Mules, 1992; Hall, 1996). Cities and countries hosting international competitions can take advantage of the media coverage of the sports events, as they may create a positive image and identity of a place and they divert visitors' attention to the hosting destination (Gibson, 2005).

Technology in Sports Events

Technological development has had a great impact on active and passive sports tourists' experience. In this section an overview is given of the innovations which are frequently manifested in sports events in our days.

First of all, active sport tourists are supported by technology in many ways during a competition depending on the types of sport, the size of the events and the organizers' dedication to technological solutions as opposed to human resources.

The Internet of Things (IoT) is rapidly evolving in many forms at a sports events. In general terms, IoT can be defined as an interconnection among devices. By the Internet of Things a network of physical objects can be established, where objects are able to communicate with each other and with their external environment as a result of the technology integrated in them (Kontogianni and Alepsis, 2020). Scoring systems, smart stadiums, smart devices, and various types of wearables (e.g., bracelets) denoting information on athletes' health and performance represent fast evolving examples for the IoT appearing at sport events.

Passive sport tourists as consumers play an active part in co-creating the experience when they start to interact with active participants of an event, athletes and their supporting personnel, as well as with other spectators of a competition. Event organizers seemingly take advantage of the benefits of technology with the aim of mediating and upgrading spectating experiences (Radzi et al., 2016). Immersive technologies, such as augmented reality (AR) have a great potential for enhancing tourism experience by the "augmentation" of the real world with audio, video, graphics, GPS data or other multimedia content. The user typically points a smartphone or any other digital device at a real object to see virtual objects bearing extra information overlaid on the real world in a digital context (Kontogianni and Alepsis, 2020). While AR combines virtual and real-world images, virtual reality (VR) is a computer-generated simulation of an environment. While VR is commonly used as the "starting point" of an experience frequently before travel, AR is rather used to supplement an already existing experience. Previous research suggests that besides improving the tourist experience, AR and VR can also be used for marketing purposes (Moro et al., 2019). Attractiveness, information generation, experience capacities, playfulness are some of the most influential factors of adopting AR by tourists, and the widespread use of mobile phones and handheld computing devices greatly contributed to AR adoption (Hassan et al, 2018).

Technological infrastructure created the grounds for digital marketing, which includes online marketing techniques, such as the use of social media, Search Engine Optimization (SEO), Pay per Click Management (PPC), branding, content marketing, video marketing, and the creation of apps. Previous research has demonstrated that social media presented opportunities for destination brand enhancement by co-branding with sport events. Neuhofer, Buhalis and Ladkin (2012) suggest that technological innovations are sources of creating an "enhanced destination experience", which also applies to sport tourism destinations.

Sport tourism markets inevitably grow by the use of social networks based on the Internet. Active social media users share information with their online contacts and they seem to rely more on the user generated content (UGC) than on professional information generated by brand marketers or tourism business in general (Morgan et al., 2021).

The development of technology and digitalization resulted in the accumulation of digitized data related to hotel reservations, transport and other tourism services. The so-called Big Data contributes to a better understanding of the consumer and may provide stakeholders of tourism with useful information to make informed decisions. Not only company information has become available but also a large set of data has emerged from user-generated content, such as text posts, reviews, online comments, shared photos and videos, etc. However, a great proportion of these data is unstructured, therefore the emergence of new forms of analysis and novel techniques has become essential.

Recognizing the growing amount of textual data led to the rise of text mining as an analytical technique (Fayos-Sola and Cooper, 2018). Text mining can be used to extract meaningful information from unstructured textual data and process them for a specified purpose (He et al., 2013). For example, based on the technique of text mining a sentiment analysis in social media may be used to identify people's emotions and opinions by capturing the number of positive and negative words in a set of textual data (Garner et al., in press). The technique "text mining" is often referred to as "text analytics" in a business context, where statistical, linguistic (e.g., Natural Language Processing), artificial intelligence and Machine learning methods are applied to structure and analyse the information for a business purpose (Khan and Afreen, 2021).

Since the emergence of the mobile phones, the widespread use of video cameras has resulted in a tremendous amount of video data, as well. Analysing these data automatically is a constantly evolving field, which is known as video content analysis or video analytics (VA). Video analytics been widely applied in sports events and in the various domains of tourism such as transport, accommodation, and catering services. In sports games the so-called Video Assistant Referee (VAR) system has gained ground, which not only monitors the game, but also analyses players' performance in real time.

Crowd analysis has become significant with the increased attention to safety and public health. Crowd management is especially important in sports halls, stadiums and other sports venues where people may gather in large numbers. Crowd behaviour analysis is built upon studying individual as well as group behaviour to determine potentially harmful behaviour, and it requires a firm technological basis including video camera systems, data mining, artificial intelligence and computer vision. Counting and density estimation, motion detection and tracking are often included in crowd analysis (Virvou et al., 2019).

A major challenge that arises here is that the data gained either from sport or tourism activities is usually that of personal nature, which on the one hand may be used for personalized experience creation, but on the other hand may lead to privacy concerns. Although all systems should comply with the EU GDPR, data protection may present difficulties in event management (Habegger et al., 2014).

Case Study of an International Sports Event: The 2021 Sabre World Cup in Budapest, Hungary

Fencing is a prestigious and highly respected competitive sport in Hungary, whose history as an organized sport dates back to the beginning of the nineteenth century in the country, when the first National Fencing Institute of Pest was founded to train young people to fight. The sport already appeared at the inaugural Olympic Games in 1896 and the three disciplines of fencing – foil, sabre and epee – have featured in every modern Olympic Games since then. Fencing is governed by the International Fencing Federation (Fédération Internationale d'Escrime-FIE) with its head office being in Lausanne, Switzerland, and it is composed of 153 affiliated national federations. The FIE establishes the rules according to which international competitions must be organized and oversees their implementation (FIE, 2021).

In Hungary fencing is the most successful Olympic sport, as 38 gold, 24 silver, and 28 bronze medals have been awarded to Hungarian fencers during the history of the Olympic Games (Olympics, 2020; Hunfencing, 2020). It is also a culturally significant sport as being proud of the good fencers can be considered as a part of the national identity.

Hungarian cities (e.g., Budapest, Debrecen, Eger, etc.) regularly host various types of fencing events, such as World Cup, European championships and national competitions, as well as small-scale fencing tournaments, for example national and regional championships for senior, veteran, youth and wheelchair fencers.

In 2021 the Hungarian Fencing Federation hosted the "Gerevich-Kovács-Kárpáti" Individual Men's and Women's Sabre World Cup and Zarándi Csaba Sabre Team World Cup between 11–14 March in Budapest, Hungary. The significance of these sports events lies in the fact that these were the last Olympic-qualifying World Cup before the 2021 Olympic Summer Games in Tokyo. Qualification in fencing is primarily based on the FIE Official Ranking, while further individual places are available at zonal qualifying tournaments, such as the above-mentioned World Cup.

This cup was also considered as a milestone in the history of modern fencing as these were the first international sporting events organized after a one-year break resulting from the COVID-19 restrictions. As these events were still organized during the pandemic, special safety and sanitary measures were taken to mitigate the risk of COVID-19 contagion as much as possible among the participants. Spectators were not admitted to the venue to watch the sports competition, therefore the technology providing opportunity to follow the events became more important than ever before. Accessibility to the events could be accomplished from any geographical location by means of digital technology.

As it is well-established in tourism literature, the tourism journey can be divided into three consecutive phases: the before travel (pre-travel), during travel (or on-site) and after travel (post-travel) periods. In the following section the characteristics of the technological environment of the above-mentioned sports events are presented in the context of these phases.

During the pre-travel period the future participants of the event were primarily informed by the website of the FIE (www.fie.org/competitions), as well as the official website of the tournament (www.gkk2021.hunfencing.hu/). The latter website was created and maintained by the Hungarian Fencing Federation, which was responsible for organizing the event. First of all, entering the competition was accomplished online, as is usual for large sports events. The website was also used for providing practical information on the venue of the events, accommodation, catering, as well as the modes of transfers between hotels and sport venues.

Due to the COVID-19 restrictions, at the time of the competition the borders of Hungary were closed except for pursuing business or economic activities, including participating in sports competitions. Therefore the participants were not allowed to arrive earlier than the start of the event, and extended stays were not permitted either, which otherwise could have provided opportunities for touristic activities, as was usual before the pandemic. Booking of the tourist services (accommodation, transfer, etc.) was also implemented by the use of the Internet, although several special arrangements were made because of the pandemic situation. The participants of the event were informed in advance that the Sabre World Cup is organized in the so-called "Bubble System", whose guidelines were carefully defined in a "COVID protocol" and disseminated through the website of the event. According to these guidelines, all 390 participants (athletes and their teams, referees, FIE officials, etc.) were staying at two designated hotels in Budapest and the participants had to produce two negative PCR test results to leave their hotel rooms, where they were obliged to self-isolate after arriving in Hungary. Even if they received the negative test results, movement was only allowed within the "bubble", which meant that the participants could only move in the specifically designated hotel where they were accommodated and at the venue of the competition. The only way to reach the venue of the event was to use the official shuttle buses or walking, otherwise participants were strictly forbidden to leave the "bubble". If a participant wished to do any shopping, for example, they were permitted to arrange it exclusively online. It must be noted that the "bubble system" created an extremely high financial burden on the organizers and was not as efficient as expected since eight international participants were recorded to produce positive PCR and antigene test results during the event in spite of the strict COVID protocols. It seems obvious that organizing this international sports event required high-quality broadband Internet connection as a fundamental technological support all through the periods of preparation and execution phases.

A separate menu of the official website was dedicated to the so-called "COVID information", where measures to control the disease were described in detail and all the necessary documents for entering the country during the pandemic were uploaded. The website also contained the entries to the competition and during the event the live stream broadcasting and the results were easily accessed online through the homepage of the World Cup, too. With the official website of the event being the most important and most reliable source of information for athletes, coaches and all the other participants, the event organizers seem to have been aware that this online platform had to be precise and up-to-date (GKK WC 2021 Budapest, 2021).

Besides gathering information on the sports event, the athletes were advised to familiarize themselves with education resources by using e-learning platforms, such as the Anti-Doping Education and Learning platform (ADEL). This e-learning platform is operated by the World Anti-Doping Agency (WADA), whose aim is to promote and coordinate anti-doping in sport internationally. ADEL is a global platform that provides education on clean sport and antidoping to athletes and their families, coaches, medical professionals, and the athlete support personnel in general by offering interactive education courses and informa-tion resources. The education programmes are supported by a wide range of interactive online tools, for example e-learning materials, podcasts, factsheets, presentations, pre-recorded webinars and live webinar series, video tutorials and online quizzes, and so on. It also encourages stakeholders of sports to connect with others involved in anti-doping (Adel, 2021).

While the webpages of the World Cup provided sufficient information on the organizational aspects of the events, marketing of the competition was primarily implemented by the use of the social media platforms. First of all, Facebook was chosen to promote the event among the supporters and the wider audience. A few days before the World Cup, the Hungarian Fencing Federation started to share new contents on its Facebook page in order to arouse interest in the competition among the public. There were short videos posted regularly which featured interviews with coaches, athletes and sports managers. On one occasion a video was shot to present an athlete's hotel room, although the focus was on the safety measures and the sportsman's personal preparation for the tournament and not the services of the hotel (www.facebook.com/hunfencing). It is worth mentioning that most of these videos were only in Hungarian language and they were not even subtitled in English, which restricted the size of the potential audience. The social media platform could also be an effective tool for creating and developing a posi-tive destination image of the city and the country hosting the sports event. However, in this case no effort could be detected to utilize this opportunity. Besides the event organizers, the International Fencing Federation also actively used its Facebook page (www.facebook.com/fie.org) to communicate with its followers and share information on the event.

During the on-site phase of sports travel, the use of technological advancements came into the foreground. On the one hand, it results from the characteristics of contemporary fencing, in which electronic equipment is common, for example electronic scoring is used to record points, foil and sabre fencers wear an elec-tronically conductive jacket, wireless fencing systems are in use, and so on. On the other hand, technology has been extensively involved in sports event organ-ization since the onset of the pandemic.

First, social media platforms were used to inform and communicate with the spectators during the days of the events. Supporters could easily follow the rounds and their results on the Facebook page of the Hungarian Fencing Federation. Posts composing of photos, videos and texts were shared multiple times a day showing the outstanding moments of the World Cup.

As spectators were not permitted to the venue of the Sabre World Cup of Budapest in 2021, high-quality live broadcasting of the events in real time became more important than ever before. Live streams of the events were shared on YouTube, which could easily be accessed via the webpage of the World Cup, as well. The live streams with professional sports commentaries provided an opportunity for spectators to experience the sports events, and to react on the different aspects of the content. Following the live streams and reacting to them contributed to building a virtual community among the members of the audience. The spectators and participants of the tournaments also joined informal Facebook groups, where they could share information and interact with each other.

It is worth noting that the sports commentaries of numerous international fencing championships, and also that of the Olympic Games, are delivered by a team of Hungarian broadcasters led by Barna Héder, the Head of the TV production at the International Fencing Federation. He has produced the commentaries on fencing at seven Olympic Games so far, and he has played a major role in improving the quality of sports broadcasting by introducing a new approach to the live coverage of fencing. He placed the spectators' experience at the centre, which implied that the most spectacular movements were focused on with the help of slow-motion reviews and instant replays. Slow-motion technology has greatly improved for the last few years and allowed the commentators to explain the delicacies of a sports match to the audience. During a fencing competition there is constant communication among the referees and the broadcasters; referees may only start a fencing bout after the broadcaster signalled to them that they have finished the slow-down and the commentaries.

Nowadays cameras are not only used for broadcasting purposes, but also for video refereeing. Video refereeing or video assistant refereeing (VAR) technology has been introduced to some sports competitions, such as football, rugby and hockey, for the last decade. For example, the use of video assistant referees in football was included in the Laws of the Game of 2018/2019 and sports event organizers have been supported with the implementation of the technology by FIFA, the governing body of football, as well. At football matches the VAR system implies that a team of three people work together to review the main referee's certain decisions by watching video replays of the relevant incidents. The team consists of the video assistant referee, his assistant and a replay operator. They are situated in a video operation room, which is equipped with several monitors offering different camera angles (FIFA, 2021).

In fencing, video bouts have one designated referee, with a second referee acting as video consultant. The video referee is located at the end of the piste and may be consulted by the referee in person. Fencers may appeal for video refereeing when they wish to ask for a review of a referee's decision during a bout.

Finally, the video referee's perspectives of the scene is often broadcast to the spectators who watch the event online (FIE, 2020). The video refereeing system was implemented at the 2021 Sabre World Cup in Budapest, Hungary, too.

Technology plays a vital part in management of the fencing competitions from the small-scale competitions to the international championships. Piste apparatuses are frequently connected to a specifically designed, licensed software which is capable of managing the competitions during a sports event. At the World Cup, the so-called "Engarde Smart" system was used, whose functions cover several domains. This system proposed competition formulas at the beginning of each round, which eliminated the lengthy time periods of preparation which was usual when executed by human efforts only. Real-time information on the results was always available due to the connection of the scoring systems and the piste apparatuses. The results were displayed on-site and on the Internet. With Engarde Smart the participants of the competitions had quick and easy access to information on their smartphone, such as their next match to fence, time, piste and opponent or their previous results. Not only the fencers, referees and the personnel on-site but also the spectators could constantly follow the progress of the competition on their smartphones, tablets or PCs.

Another fast-developing area of technological applications is that of the health and performance analysis of the athletes. It is becoming widespread to use various monitoring systems to collect data for understanding physiological changes during a match in order to maximize an athlete's improvement in performance. However, this set of data is not public and their analysis normally takes place after the event.

In the post-trip phase of a sports travel, not only an assessment of the athletes' performance takes place but also an evaluation of the sporting event. Event organizers usually distribute online questionnaires to the participants to investigate their satisfaction with the various aspects of the event. The organizers of the Sabre World Cup in 2021 also executed an online participant satisfaction survey, which was analysed after the event. The homepage of the World Cup remained accessible after the events, as well as the recordings of the competitions are available on the online video-sharing platform YouTube.

Conclusion

This chapter has outlined the growing importance of technological innovations in sports tourism, and utilized a case study of the Sabre World Cup 2021 organized in Budapest, Hungary to illustrate some of the salient issues. Understanding the various forms of technological applications is of vital importance to sports event managers so that they can incorporate the most efficient methods and techniques to the event management process. As there is a high demand for following a sports event in an effective, personalized and entertaining way, technological innovations will probably receive an even greater emphasis in the future. In addition, tourism destination marketers should also be aware that social media driven by digital technology has dramatically changed event and destination marketing, and it has also presented opportunities for destination brand enhancement. Finally, the relationship between technological innovations and the level of tourist experience deserves more academic attention.

References

Adel (2021). *The global Anti-Doping Education and Learning platform.* Retrieved from: https://adel.wada-ama.org/learn (accessed 11 September 2021).

Burgan, B., and Mules, T. (1992). Economic impact of sporting events. *Annals of Tourism Research*, 19(4), 700–710.

Chernushenko, D. (1996). Sports tourism goes sustainable: The Lillehammer experience. *Vision in Leisure and Business*, 15, 65–73.

Csobán, K.V., and Serra, G. (2014). The role of small-scale sport events in developing sustainable sport tourism: A case study of fencing. *APSTRACT: Applied Studies in Agribusiness and Commerce*, 4, 17–22.

Fayos-Solà, E., and Cooper, C. (2018) (Eds.) *The Future of Tourism: Innovation and Sustainability.* Cham: Springer.

FIE (2020). *Olympic Games 2020 TOKYO Refereeing.* Retrieved from: https://tokyo2020.fie.org/rules/refereeing (accessed 20 September 2021).

FIFA (2021). *Video Assistant Referee (VAR) Technology Standards.* Retrieved from: www.fifa.com/technical/football-technology/standards/video-assistant-referee (accessed 10 September 2021).

Gammon, S., and Robinson, T. (2003). Sport and tourism: A conceptual framework. *Journal of Sport & Tourism*, 8, 21–26.

Garner, B., Thornton, C., Pawluk, A.L., Cortez, R.M., Johnston, W. and Ayala, C. (in press). Utilizing text-mining to explore consumer happiness within tourism destinations. *Journal of Business Research.*

Gibson, H.J. (2005). Towards an understanding of why sport tourists do what they do. In H.J. Gibson (Ed.), *Sport Tourism: Theory and Concepts.* Abingdon: Routledge, pp. 66–85.

Gibson, H.J., Kaplanidou, K., and Kang, S.J. (2012). Small-scale event sport tourism: A case study in sustainable tourism. *Sport Management Review*, 15(2) 160–170.

GKK WC 2021 Budapest (2021). *Home.* Retrieved from: www.gkk2021.hunfencing.hu (accessed 11 September 2021).

Habegger, B., Hasan, O., Brunie, L., Bennani, N., Kosch, H., and Damiani, E. (2014). Personalization vs. Privacy in big data analysis. *International Journal of Big Data*, 25–35.

Hall, C.M. (1996). Hallmark Events and Urban Reimaging Strategies. Coercion, Community and the Sydney 2000 Olympics. In L.C. Harrison and W. Husbands (Eds.), *Practising Responsible Tourism: International Case Studies in Tourism, Planning, Policy, and Development.* New York: Wiley, pp. 336–379.

Hassan, A., Ekiz, E., Dadwal, S.S., and Lancaster, G. (2018). Augmented reality adoption by tourism product and service consumers: Some empirical findings. In T. Jung and M. tom Dieck (Eds.), *Augmented Reality and Virtual Reality.* Cham: Springer, pp. 47–64.

He, W., Zha, S., and Li, L. (2013). Social media competitive analysis and text mining: A case study in the pizza industry. *International Journal of Information Management*, 33(3), 464–472.

Hiller, H. (2006). Post-event outcomes and the post-modern turn: The Olympics and urban transformation. *European Sport Management Quarterly*, 6, 317–332.

Hinch, T.D., and Higham, J.E.S. (2001). Sport tourism: A framework for research. *The International Journal of Tourism Research*, 3(1) 45–58.

Hinch, T., and Higham, J. (2011). *Sport Tourism Development* (2nd ed.). Wallingford: Channel View Publications.

Hunfencing (2020). *Archív eredmények*. Retrieved from: https://hunfencing.hu/archiv-eredmenyek (accessed 15 September 2021).

Kaplanidou, K., and Gibson, H.J. (2010). Predicting behavioral intentions of active event sport tourists: The case of a small-scale recurring sports event. *Journal of Sport and Tourism*, 15(2), 163–179.

Khan, A.M., and Afreen, K.R. (2021). An approach to text analytics and text mining in multilingual natural language processing. *Materials Today: Proceedings, 2021*, DOI: 10.1016/j.matpr.2020.10.861.

Kontogianni, A., and Alepis, E. (2020). Smart tourism: State of the art and literature review for the last six years. *Array*, 6, 100020.

Morgan, A., Wilk, V., Sibson, R., and Willson, G. (2021). Sport event and destination co-branding: Analysis of social media sentiment in an international, professional sport event crisis. *Tourism Management Perspectives*, 39, 100848.

Moro, S., Rita, P., Ramos, P., and Esmerado, J. (2019). Analysing recent augmented and virtual reality developments in tourism. *Journal of Hospitality and Tourism Technology*, 10(4), 571–586.

Neuhofer, B., Buhalis, D., and Ladkin, A. (2012). Conceptualising technology enhanced destination experiences. *Journal of Destination Marketing & Management*, 1(1–2), 36–46.

Olympics (2020). *Tokyo 2020 Medal Table*. Retrieved from: https://olympics.com/tokyo-2020/olympic-games/en/results/all-sports/medal-standings.htm (accessed 11 September 2021).

Radzi, S.M, Hafiz, M., Hanafiah, M., Sumarjan, N., Mohi, Z., Sukyadi, D., et al. (2016) (Eds.). *Heritage, Culture and Society, Research agenda and best practices in the hospitality and tourism industry*. Abingdon: CRC Press.

Virvou, M., Alepis, E., Tsihrintzis, G.A., and Jain, L.C. (2019) (Eds.) *Machine Learning Paradigms: Application of Learning and Analytics in Intelligent Systems*. Cham: Springer.

12 Technology and Events
The Case of Note di Fuoco Festival in Calabria in South Italy

Debora Calomino

Introduction

The relationship between events and advanced technology in recent years has increasingly strengthened. Cultural and sporting events are enveloped in the most diverse technologies, both for the organization of the event and for the management of visitors, and to make the experience spectacular and inclusive. In the era of experiential tourism, the use of innovative tools is an added value that allows the individual event to become unforgettable. The involvement of the five senses, the possibility of attending certain events even at great distances thanks to live streaming, the use of drones, holograms, technologically advanced booking platforms, the diffusion of images in real time on social networks that allow you to reach users all over the world, are just some of the elements put in place to improve the enjoyment of the event. This chapter analyses a cultural event dedicated to the pyrotechnic art that takes place in Calabria, a region of Southern Italy. The event is called Note di Fuoco and numerous technologies are used to make it happen. From the dedicated app for logistics management, to the detection of polluting elements, passing through the promotion on social channels, the whole event is the result of the most modern technologies. It would be impossible to do it without them.

Technology and Events

Events have a strong impact on the development of territories. They are very important for destinations' competitiveness for many reasons. First of all, they are tourist resources that increase local offering, frequently enhancing other local assets. In addition, they attract new and different tourist segments and reduce seasonality. In fact, the events have a limited duration in time, are unique phenomena and much of the appeal of events is that they are never the same and you have to be there to enjoy the unique experience fully; if you miss it, it's a lost opportunity (Getz, 2008). Events also have a strong impact on the image of the destination. The image can be defined as a set of beliefs, ideas and impressions of a person about an object, that is, a mental construction, based on numerous elements, which influences the perception of the object itself (Kotler, 2001). One

DOI: 10.4324/9781003271147-17

of the most relevant aspects of the impact of an event on the territory, especially if it is a tourist destination, is related to the effects of the event on the image and on the tourist brand of the locality (Ferrari, 2018).

In recent years technologies have given a big help to the spread of events: the possibility to buy and sell tickets online, the sharing of promotional videos, the use of websites in which to find a lot of information to reach the destination of the event, details of accommodation and other attractions. Also social media contribute to make an event an experience not to be missed. Events have evolved over time thanks to technology. The aim is to make the experience complete, both for the organizers and for the users. According to Pine and Gilmore (1998), today consumers want to live memorable experiences; in fact we talk about the economy of experience, precisely to put the concept of experience at the centre.

From an organizational point of view, technology helps to make the process more fluid, thanks to a series of assistance, detection and flow management tools. From the point of view of the user, technology has improved the experience, increasing entertainment and the possibility of enjoying the event itself. For example, augmented reality viewers make the experience more complete, thanks to additional content, which is difficult to disclose without the help of technology. Augmented reality also improves emotional involvement with visitors, allowing them to experience a unique and unrepeatable experience. Applications serve both an informational and an organizational purpose. Chatbots are virtual assistants that help guests find basic information quickly and efficiently. Thanks to the chatbots it is possible to give information on attractions in the surroundings of the event location or practical information on how to move nearby. Another tool available for events are beacons, low-power radio transmitters that use Bluetooth technology to detect the presence of smartphones and send information to them. During an event they can be useful for detecting attendance, accrediting guests, offering assistance to users, sending information material, requesting feedback and giving promotional coupons. The advantage of this technology is that it is not invasive, in fact it is not necessary to download any app to use the services. Live streaming allows you to reach a greater number of audiences, broadcasting the event live on dedicated platforms. In the pandemic period, streaming made it possible to digitize events, virtualizing the physical space. Specifically, technology has been essential to organize webinars, online meetings, digital daily events, virtual fairs.

Gaming is often used during events to increase the involvement of the participants. Inserting playful elements for the enhancement of an event allows you to create insights, increasing attention to the contents of the event. This expedient enhances the event, especially if it is aimed at a youth target who can also benefit from the learning point of view.

In order to film events from multiple points of view, the use of drones (remotely piloted aircraft) is increasingly widespread. This particular technology allows you to have an all-round view. In addition, the shots taken with drones are useful for creating virtual tours of the places, to familiarize yourself with the venue where the event will take place, before its realization.

Case Study: Note di Fuoco

Note di Fuoco is a cultural event dedicated to the art of pyrotechnics. It is classified as a hallmark event, that is a recurring event linked to a locality, which has elements of attraction (tradition, image, notoriety) such as to increase the competitiveness of the host places and become a specific characteristic, as well as an element of differentiation from the point of view of marketing (Ferrari, 2018). Hallmark events appeal to both local and external audiences (Nicholls et al., 1992). According to Note di Fuoco (n.d.), it was conceived by the CreativaMente social promotion association, founded in 2012. During the event, Italian and foreign bomb disposal masters perform, giving life to pyrotechnic shows in time to music, staging a different theme every night. The event has grown more and more over the years, attracting both Italian and foreign visitors and tourists from Spain, Austria, Poland, Luxembourg, England, Germany, Czech Republic, Belgium, and Holland. The event lasts five days, during which the mastery of the pyrotechnic artisans blends with the most advanced technologies. The fireworks are placed on floating rafts, 400 metres away from the spectators and this allows a wide overview that allows you to enjoy the show, even from a long distance. The show is divided into several levels, making the event a unique event of its kind in southern Italy. Not only fireworks but also concerts, guided tours, street artists, fashion shows and sports competitions. Until 2019, 840,000 admissions were registered, 10,000 tickets were sold every evening, with 300 artists participating in the event and 500 people working for the success of the event. Note di Fuoco is included in the list of historicized events of the Calabria Region, therefore considered one of the most important events in the Calabrian territory, capable of attracting thousands of visitors, and it is strategic from a tourist point of view.

Technology and Note di Fuoco

Again, according to Note di Fuoco (n.d.), technology is a key element for the success of Note di Fuoco; in fact the whole event is supported by it. Starting from the website www.notediFuoco.it which collects all the information relating to the event, divided into sections for easy consultation, the event also made use of a dedicated application, for the logistic management of the event, the flow of visitors, the management of available parking spaces and the use of shuttle buses. Social media, in particular Facebook, Instagram and YouTube, were of fundamental importance for the promotion of the event. Real-time updates on scheduled events and a series of useful information for visitors have been published on Facebook and Instagram. The live streaming of the highlights of the event was broadcast on Facebook and YouTube, thanks to the live streaming made available by the platforms. At the entrances of the area dedicated to the event (an area of 18,000 square metres) a system of cameras was placed to control the entrance and exit of visitors, monitored by a mobile station with a dedicated server, which every ten minutes measured the real number of spectators, as the limit was set at 10,000 visitors for safety reasons. For safety reasons, drones were also used, which flew over the area every 20 minutes, thanks to two operators. Note di Fuoco also uses technology to protect the environment: control units have been installed to monitor the emission of fine dust, detecting the index both during the shooting and afterwards, in order to keep the environmental impact of the event. After the event, thanks to the support of specialized divers, it is possible to clean up the seabed, to eliminate paper residues and other materials used for the success of the fireworks.

Conclusion

The analysis of the Note di Fuoco event dedicated to the art of pyrotechnics in Calabria, Southern Italy, highlighted how much the use of the most advanced technology makes it possible to create the experiences related to the events more complete and inclusive. Thanks to technology, the experience of the event is improved: all the senses are involved, not just sight, which has always been considered a privileged sense. Hearing and touch in particular are involved, especially during events in which 3D viewers are used that allow immersive experiences. Note di Fuoco is an event that uses technology for various aspects: both to improve the user experience and to help the organizers to better manage the event. In the future, the study of further technologies applied to the realization and use of events will make these experiences even more unforgettable. The aim is to study new methods of valorizing territorial resources through increasingly technologically advanced events. Improving the fruition and quality of events will be one of the goals of technology in the coming years. Better manage the flow of visitors, increase the reputation of the event, enjoy the events also on the other side of the world, thanks to increasingly advanced technologies that will make it possible to approach places and people who are physically very distant.

References

Ferrari, S. (2018). *Event Marketing- I grandi eventi e gli eventi speciali come strumenti di marketing*. Milan: CEDAM.

Getz, D. (2008). Event tourism: Definition, evolution and research. *Tourism Management*, 29, 403–428.

Kotler, P. (2001). *A Framework for Marketing Management*. Upper Saddle River, NJ: Prentice-Hall.

Nicholls, J., Laskey, H., and Roslow, S. (1992). A comparison of audiences at selected hallmark events in the United States. *International Journal of Advertising*, 11(3), 215–225.

Note di Fuoco (n.d.). *Home*. Retrieved from: www.notedifuoco.it/ (accessed 1 September 2021).

Pine II, B.J., and Gilmore, J.H. (1998). Welcome to the experience economy. *Harvard Business Review*. Retrieved from: https://hbr.org/1998/07/welcome-to-the-experience-economy (accessed 1 September 2021).

13 The Event "7 Wonders of Gastronomy" and the Digitalization of Communication in the Portuguese Context

Bruno Sousa and Beatriz Casais

Introduction

Currently, globalization and significant competition between territories (i.e., cities, regions or countries) lead to new challenges in terms of territorial management and local development. Territories and tourism destinations have always felt the need to differentiate themselves from others and to affirm their uniqueness (distinctive in their own characteristics). Consequently, the heritage of wine and gastronomy is seen as a differentiating factor from the perspective of residents and tourists. Gastronomy and wine are decisive for the success of tourism and local development, both for the satisfaction of visitors and the desire to return in the future and recommend to family and friends. Throughout history, food has played an essential role in nutrition being the way in which individuals satisfied their most primitive need.

According to Gross and Brown (2008), tourism knowledge has advanced by applying theories developed in other disciplines and so it is surprising that some research streams have been developed, tested and widely reported in leisure journals without attracting the interest of tourism researchers. This is characteristic of the situation that exists with regard to research about involvement, place attachment and, more recently, the combined use of several constructs.

In this sense, place attachment attracted researchers' interest when investigating the relationship between psychological impressions of people with aspects related to geography, architecture, environment, tourism or leisure. Ultimately, place attachment becomes instrumental in supporting the management of destinations and territorial planning (Green and Chalip, 1997; Hwang et al., 2005; Poço and Casais, 2019; Sousa and Rocha, 2019). Based on relevant inputs from previous studies (e.g., Ramkissoon et al., 2013; Yoon and Uysal, 2005), and an exploratory study, we advance research propositions that expand previous studies developed in similar contexts. The study further allows to draw relevant managerial implications and suggestions for future research in specific contexts of food and wine tourism (e.g., "7 Maravilhas da Gastronomia").

DOI: 10.4324/9781003271147-18

Place Attachment and Food and Wine

The concept of place attachment presented, from the beginning, some difficulties in the way it was defined or understood universally, considering the most appropriate methodological approach. Other terms are used in an undifferentiated way such as community attachment (Kasarda and Janowitz, 1974), community sense (Sarason, 1974), local identity (Proshansky, 1978), place dependency (Stokols and Shumaker, 1981), and sense of place (Hummon, 1992). The place attachment to the country would be key for creating loyalty among the country's citizens and for them to recommend the country to foreign visitors (Gross and Brown, 2008; Casais and Sousa, 2019). It would also have a social and economic impact by increasing tourism and would lead to a value co-creation of a national brand (Shams and Lombardi, 2016). For instance, and according to Santos et al. (2021), tourist involvement is an emotional state of interest or enthusiasm in relation to place or product, which ends up influencing an experience in the destination. Food or wine involvement is essential for the food and wine lover to decide to visit a certain region, in order to have a fuller experience of immersion in the product or winescape, not only through the tasting, but also through all activities related to product culture.

The very particular nature of food and wine tourism depends on sensory involvement, through all the tangible and intangible aspects of product or winescape (Brochado et al., 2021; Santos et al., 2021).

In certain contexts, it is found that one of the terms used as a generic concept encompasses some dimensions, serving as an example (Lalli, 1992; Tongue et al., 2015), the attachment to a place is a component of identity. On the other hand, some authors use them in distinction as if they were synonyms such as Adams (2016), showing attachment and identification separately. In fact, not only does it recognize the existence of an affective bond with the places, but also the importance that this can have in positively or negatively qualifying existence. And not only individual existence, private, but also the existence of human groups (Sousa and Rocha, 2019).

According to Poço and Casais (2019: 226), the "competitiveness of places has been growing with no signs of slowing down. People are more informed and demanding when looking for a place to live or visit and the development of an emotional marketing has been an important dimension in city and tourism management. The literature has focused on factors promoting place attachment and its consequence in place satisfaction and loyalty." It is natural for people to develop feelings or emotions, negative or positive, pleasant or unpleasant, for places related to past or current experiences; for example, places connected with childhood, or linked to future perspectives (such as the dream place to live or to go/come back) (Sousa and Rocha, 2019). Perhaps the absence of a feeling of mutual affinity, community, fraternity among people, in a formal or informal way, institutionalized or not – no sense of diversity, aversion, and hostility – is in a way related to issues of place, territory, and attachment to places (Lee, 2001). According to Poço and Casais (2019: 227), the "emotional place attachment, has attracted significant attention in the study of individuals' bonds to a specific place. Emotional attachment to the place implies the attribution of emotions and that represents the perceived relationship of an individual with functional dependence to a place. For individuals create emotional bonds with places by developing regular contacts, with specific contexts, over more or less lasting periods".

Therefore, according to Gross and Brown (2008: 1143), "the use of the place attachment and involvement constructs in combination has occurred only recently in leisure studies, and in the context of recreation. The pioneers in this area have been Kyle and his co-scholars, who studied involvement (e.g., food and wine) (Kerstetter et al., 2001; Kyle and Chick, 2002; Scott and Shafer, 2001) and place attachment (Kyle et al., 2003a; Moore and Graefe, 1994) on separate research tracks until combining them in a key 2003 study that measured the relationships between leisure activity involvement" and place attachment among hikers on the Appalachian Trail (Kyle et al., 2003b). In this context, scholars have assessed the social significance of food (Goody, 1982; Cook and Crang, 1996; Higgins-Desbiolles, 2006), the role of food in tourism (Bessière, 1998; Hall and Mitchell, 2005, 2000; Hjalager and Corigliano, 2000; Hjalager and Richards, 2003), the semiotics and symbolism associated with food and also related topics such as the effects of globalization on food (Hall and Mitchell, 2000), thereby confirming that food does have a role to play other than simply as a means of nourishment, and as a contributing factor to the enhancement of tourists' experiences (Chaney and Ryan, 2012).

Therefore, and according to Björk and Kauppinen-Räisänen (2014), contemporary consumers are increasingly showing their interest in local food. These products are bought as daily purchases at local retailers and supermarkets, and shopped for in farm shops, food cooperatives and farmers' markets. Food cooked from local ingredients is also served at restaurants as a means of adding value to their patrons' eating experiences and to respond to customers' recently acquired interest in local food markets (Pieniak et al., 2009). Cultural tourism plays a major role in the economic development of many cities (Covas, 2009; Eusébio and João Carneiro, 2012; Ribeiro et al., 2012; Vareiro et al., 2012). In this context, gastronomy has the ability to convey a sense of the heritage and cultural identity of the host communities and therefore the authenticity of the experience (Chaney and Ryan, 2012), and the ability to convey prestige, status and to create groups by inclusion or exclusion (Getz, 2008; Moscardo, 1999; Chaney and Ryan, 2012). Food and wine and cuisine are one means of product differentiation in a tourist market where destinations that may be the same in terms of retail provision, buildings, climates and other characteristics vie for the tourist expenditure. The positive relationship between gastronomy and tourist activity is an economic catalyst for growth in specific territories, promoting local products and providing added value to the tourist. In this context, gastronomic tourism is an excellent business opportunity for Portugal if the country is able to capitalize on the development of high-quality agriculture and food, especially related to the election of the "7 Wonders of Gastronomy".

The "7 Wonders of Gastronomy" in Portugal

After divulging and promoting the historical and natural heritage of Portugal, the year 2011 was dedicated to gastronomy, one of the great values and passion of the Portuguese. The election of the "7 Wonders of Gastronomy" divulged and promoted the national gastronomic heritage, recognized and appreciated all over the world for its diversity, unique flavours and quality of the products with which the dishes are made. The culinary arts constitute an intangible heritage, a testimony of our cultural identity, and are a decisive factor in the choice of Portugal as a tourist destination. The Portuguese recipe was promoted and safeguarded, guaranteeing its genuine character, promoting agricultural products of superior quality and privileging the regional diversity. In this section of our manuscript, we have illustrated some of the typical examples of the 7 gastronomic marvels in Portugal, which are a notorious identity card and tourist differentiator at regional and local level.

As a first example, Pudim Abade de Priscos is a typical pudding from Braga, being one of the few recipes that Abade de Priscos transmitted to the public. The pudding became known when Pereira Júnior, the director of the Primary Magisterium of Braga in the former Convent of the Congregados, asked the Abbot of Priscos for recipes to teach in the magisterium.

According to "7 Wonders of Gastronomy", the pudding is made in a brass or copper pot where half a litre of water is placed. When it is boiling, add half a pound of sugar, a lemon peel, a cinnamon stick and 50g of lard (it is proposed that it be fat and preferably from Chaves or Melgaço) and let it boil until it reaches Spartan point. Beat 15 egg yolks and mix them with a goblet of Port wine until it is half full, after beating again. The sugar syrup is then poured through a thin colander into a bowl where the yolks are stirring, stirring everything. A sugar-sweetened form is poured into the caramel and the preparation is then poured into a water bath for 30 minutes. The pudding is unformed when it is almost cold.

As a second example, we present Leitão da Bairrada. Since the time of the Romans the roast suckling pig is known. However, the birth record evidenced by oral and written testimonies dates back to the 1920s. The first places of sale were the hotels of all the region, in which great roasts were born. Approximately 3000 piglets a day are made and the suckling pig is served in dozens of restaurants and prepared by hundreds of roasters in their homes, bringing thousands of lovers to Bairrada throughout the year. According to "7 Wonders of Gastronomy", it is seasoned with garlic, salt and pepper. In the Bairradino oven it is heated with firewood – vines or eucalyptus – and the piglet will then roast slowly, for about two hours, "stuck" on a stick. Its quality is due to the choice of breeds, such as Bisara (original and preferred breed) and the various breed crosses that currently exist. It should be cut into small pieces, without being superimposed, served on platters with skin always facing up, and accompanied by small potatoes cooked with skin, orange and salad. The roasted suckling pig is a gastronomic product recognized by the Regional Tourism Authority of Portugal.

As a third example, we present Pastel de Belém. At the beginning of the nineteenth century, in Belém, near the Jerónimos Monastery, it worked a refining of sugar cane associated with a small place of varied commerce. As a consequence of the liberal revolution in 1820, all convents in Portugal were closed in 1834, and the clergy and workers were expelled. In an attempt to survive, someone from the Monastery puts up for sale at this store some sweet pastries, soon called the "Pastéis de Belém". At the time, the zone of Belém was distant from the city of Lisbon and the route was assured by steamboats. However, the grandeur of the Jerónimos Monastery and the Torre de Belém attracted visitors who quickly became accustomed to savouring the delicious pastries from the Monastery. In 1837, the "Pastéis de Belém" began to be manufactured in facilities adjacent to refining, according to the old "secret recipe" from the convent, which is exclusively known to the master masons who manufacture them in the "Oficina do Segredo". This recipe stays the same until today. In fact, the only true factory of the "Pastéis de Belém" is able, through a careful choice of ingredients, to provide today the palate of the old Portuguese sweets.

As a fourth example, we present Açorda à Alentejana. According to "7 Wonders of Gastronomy", there is only one, "the classic", the traditional "açorda de garlic". The lunch usually consists of açorda with olives. With the classic açorda Alentejana, whose broth the barbecue prepares in an instant, the water boils over the barrels, where the cook deposits the seasonings – olive oil and salt minced with garlic, poejos or coriander and pepper. Then each one pulls on the razor and everyone starts mending the bread to the baskets, until it no longer fits him. In açorda à Alentejana, the "açorda de garlic", the ingredients are coriander and/or garlic and salt poejos; olive oil, water and bread. Break the coriander and the poejos, chop the garlic and step with salt in a mortar. Add the mixture and the olive oil to a bowl and combine with a spoon. Boil water where they can boil eggs (or poach), hake or cod to accompany, and lie in the bowl with the mixture. Finally sliced hard bread soups are laid.

Digital Communication and Territorial Market Strategies

Social media marketing (SMM) aims to produce content that users share in their various social media applications in order to increase brand exposure and broaden customer reach. There are numerous marketing techniques to apply in social media in order to involve the customer, some of which have costs and others do not. Digitalization was a real challenge for any hotel company, requiring cautious and well-planned action to be successful (Sousa et al., 2021). According to Casais and Sousa (2019), nowadays, research defines digital engagement as an important asset of brand equity, which can be measured based on user interactions in digital media – for instance, web traffic, number of clicks or page views and number of likes, comments or shares. The Internet has changed the way consumers engage with brands and much of that change comes from social media, which are changing the way companies communicate with consumers and increasing the performance of companies with these innovative channels of communication.

The digital media buzz created through the viral sharing of the challenge for its humorous style, the celebrity impact and the social responsibility dimension of the project resulted in several media companies quickly looking to be partners or media sponsors and increasing the project's visibility in mass media free of charge (Casais and Sousa, 2019). Therefore, several studies in marketing and tourism contexts have discussed the association between affection to the site and consumer purchasing behaviour, including the study of satisfaction, loyalty or quality of service (Hwang et al., 2005). In specific contexts of tourism, as food and wine tourism, there are several challenges to tourism research level (Gheorghe et al., 2014), as the relationships between affection of a particular place and behavioural intentions by the visitor. The literature does not show convincing results with respect to the effect of the local affectivity satisfaction (Yüksel et al., 2010), as already had been mentioned earlier. Some authors argue that satisfaction, depending on how this is defined, can positively influence local affection (Hou et al., 2005). However, a considerable group of authors argues that consumer satisfaction, particularly in tourism contexts, may be influenced by the local affection type (Halpenny, 2010; Yüksel et al., 2010).

It is well established in the literature that the tourist's assessment of different destination attributes influences his or her overall satisfaction and, subsequently, intentions to revisit a destination (e.g., Baker and Crompton, 2000; Kozak, 2003; Alegre and Garau, 2010; Krešić et al., 2013). Respective studies typically apply linear modelling techniques in analysing these influences, such as multiple regression analysis or structural equations modelling (Krešić et al., 2013). Both literature and the ethnographic experience described earlier support the propositions presented in this chapter. References to place identity and place dependence were observed in the gastronomic forums, as being part of the satisfaction with the experience. The development of the gastronomic wonder contest described earlier allows the observation of the affection to the gastronomic products with the sharing of experiences among participants and observed the natural generosity and helping "identity" and "hospitality" card between them which may increase and reinforce the loyalty to the destination.

Discussion and Recommendations

This exploratory research allowed the perception from a marketing perspective that feelings of place identity, in the emotional dimension of the "7 Wonders of Gastronomy" contest, and the satisfaction with the experience may increase and consequently it can reinforce the loyalty to the destination. This idea was backed up by the testimonials that the researchers collected from different forums, contest participation and social media analysis and debates. The Internet has changed the way consumers engage with brands and much of that change comes from social media, which are changing the way companies communicate with consumers and increasing the performance of companies with these innovative channels of communication. Also, the place dependence of locals, companies and public institutions actively creates an integrated tourist experience. Such experience is organized and structured based on several dimensions: cultural, natural, gastronomic and commercial. These contribute to tourist satisfaction and loyalty. This study is a preliminary contribution towards a greater understanding regarding the relationship between the gastronomic products affection and behavioural intentions in specific contexts of tourism, in particular the case of the "7 Wonders of Gastronomy". As an interdisciplinary approach, this research contributes positively to the development of theory in marketing and tourism contexts. Digital communication progressively assumes itself as a fundamental tool for territorial management and planning of tourist destinations. Digital communication takes on special prominence in the (post) pandemic context, considering the changes in the population's lifestyles. In our preliminary study, it seems to be evident that digital communication assumes a significant importance as an element of territorial promotion in tourist segmentation (gastronomy and wines) in the Portuguese context (i.e., images, website, competition, television advertisements, social networks, technological sharing platforms of information). Several models might be designed in the future, connecting tourist destinations in the context of food and wine with tourist consumer behaviour, particularly focusing on aspects that reinforce brand attachment to the destination place, as planning spaces, communication strategies, promotion services, integrated experiences and combating seasonality.

Conclusion and Future Research

This study is an exploratory research based on a qualitative approach. The exploratory study of the researchers allowed them to understand the phenomena of place identity and place dependence with the participants and the consequent effect on satisfaction and loyalty. These conclusions are based on the perception of researchers derived from the preliminary research. Understanding consumer behaviour in specific contexts of gastronomic tourism will allow, among other things, gathering and providing useful information for planning of tourist destinations, as well as providing decision-making support for other stakeholders. This manuscript discusses branding gastronomic products and places, and particularly "7 Wonders of Gastronomy" in Portugal, which became a lifestyle trend not only for gastronomic motivations but especially connected to modern lifestyles, which tend to give increasing importance to identity. The aim of this study was to provide an empirical contribution in this area by presenting the "7 Wonders of Gastronomy" in Portugal. The brand identity should involve stakeholders who depend on destination and co-create the gastronomic brand place. Experiences are the result of situations that provide sensory, emotional, cognitive, behavioural, relational and functional stimuli that trigger a constant flow of fantasies, feelings and fun (Pina and Dias, 2021). Food and wine tourism allows experiences that involve the senses and emotions and provide pleasure to food and wine lovers in a rural setting. Stimulating the senses is therefore strategic in involving visitors emotionally with the products or wines of the region and its landscape, cultural and heritage context (Santos et al., 2021).

With the insights from this study it is expected that future work can contribute to the development of empirical studies to test the effect of place attachment on satisfaction and loyalty of participants in a quantitative research. An empirical study with questionnaires applied to participants should enable greater understanding of the research model and the resulting relationship among the constructs. Overall the idea would be to gauge the impact of certain factors in the predisposition to food and wine contexts and specific gastronomic tourism contexts (e.g. "7 Wonders of Gastronomy"). Thus, an empirical study in the future may measure the effect of social media promotion of the 7 gastronomic wonders with a measured place attachment of the territorial places they represent. The event is an interesting tool for disseminating the Portuguese community's gastronomy, sweets, customs, traditions and values. Some typical regional sweets have a strong reputation. The image of tourist destinations is related to local gastronomy and sweets. The event of the 7 Wonders can, consequently, enhance the affective bonds and the sense of belonging of the community. During the event, the contestants are strongly supported by the inhabitants of each region (town or city) and reinforce the (healthy) competition between neighbouring territorial spaces. The sense of belonging of the local community is sometimes reinforced and the territorial image is influenced by stakeholders. Future research may also look for a deeper understanding of the phenomenon in various gastronomic products, as well as in other locals of gastronomic tourism, in order to find differences between the place attachment to gastronomic locals and the loyalty of participants and residents. Future research should reveal several potentials of individual destination attributes to cause satisfaction and/or dissatisfaction, and shed light on the most determinant and critical attributes in explaining the overall tourist experience, in specific the case of gastronomic tourism and local development.

References

Adams, H. (2016). Why populations persist: Mobility, place attachment and climate change. *Population and Environment*, 37(4), 429–448.

Alegre, J., and Garau, J. (2010). Tourist satisfaction and dissatisfaction. *Annals of Tourism Research*, 37(1), 52–73.

Baker, D.A., and Crompton, J.L. (2000). Quality, satisfaction and behavioral intentions. *Annals of Tourism Research*, 27(3), 785–804.

Bessière, J. (1998). Local development and heritage: Traditional food and cuisine as tourist attractions in rural areas. *Sociologia Ruralis*, 38(1), 21–34.

Björk, P., and Kauppinen-Räisänen, H. (2014). Culinary-gastronomic tourism: A search for local food experiences. *Nutrition and Food Science*, 44(4), 294–309.

Brochado, A., Stoleriu, O., and Lupu, C. (2021). Wine tourism: A multisensory experience. *Current Issues in Tourism*, 24(5), 597–615.

Casais, B., and Sousa, B. (2019). "Portugal, the best destination": The case study of a CSR communication that changed mentalities and increased business performance. *World Review of Entrepreneurship, Management and Sustainable Development*, 15(1/2), 29–41.

Chaney, S., and Ryan, C. (2012). Analyzing the evolution of Singapore's World Gourmet Summit: An example of gastronomic tourism. *International journal of hospitality Management*, 31(2), 309–318.

Cook, I., and Crang, P. (1996). The world on a plate: Culinary culture, displacement and geographical knowledges. *Journal of Material Culture*, 1(2), 131–153.

Covas, A. (2009). Glocalização, reterritorialização e transformação da paisagem agro-rural: algumas reflexões a propósito. *Revista Portuguesa de Estudos Regionais*, 20, 7–11.

Eusébio, C., and Carneiro, M.J. (2012). Impactos socioculturais do turismo em destinos urbanos. *Revista Portuguesa de Estudos Regionais*, 30, 65–75.

Getz, D. (2008). Event tourism: Definition, evolution, and research. *Tourism Management*, 29(3), 403–428.

Gheorghe, G., Tudorache, P., and Nistoreanu, P. (2014). Gastronomic tourism, a new trend for contemporary tourism. *Cactus Tourism Journal*, 9(1), 12–21.

Goody, J. (1982). *Cooking, Cuisine and Class: A Study in Comparative Sociology*. Cambridge: Cambridge University Press.

Green, B.C., and Chalip, L. (1997). Enduring involvement in youth soccer: The socialization of parent and child. *Journal of Leisure Research*, 29(1), 61–77.

Gross, M.J., and Brown, G. (2008). An empirical structural model of tourists and places: Progressing involvement and place attachment into tourism. *Tourism management*, 29(6), 1141–1151.

Hall, C.M., and Mitchell, R. (2000). Wine tourism in the Mediterranean: A tool for restructuring and development. *Thunderbird International Business Review*, 42(4), 445–465.

Hall, C.M., and Mitchell, R. (2005). Gastronomic tourism: Comparing food and wine tourism experiences. In M. Novelli (Ed.), *Niche Tourism, Contemporary Issues, Trends and Cases*. Abingdon: Routledge, pp. 73–88.

Halpenny, E. (2010). Pro-environmental behaviors and park visitors: The effect of place attachment. *Journal of Environmental Psychology*, 30(4), 409–421.

Higgins-Desbiolles, F. (2006). More than an "industry": The forgotten power of tourism as a social force. *Tourism Management*, 27(6), 1192–1208.

Hjalager, A.M., and Corigliano, M.A. (2000). Food for tourists: Determinants of an image. *International Journal of Tourism Research*, 2(4), 281–293.

Hjalager, A.M., and Richards, G. (2003). (Eds.). *Tourism and Gastronomy*. London: Routledge.

Hou, J., Lin, C., and Morais, D. (2005). Antecedents of attachment to a cultural tourism destination: The case of Hakka and non-Hakka Taiwanese visitors to Pei-Pu, Taiwan. *Journal of Travel Research*, 44, 221–233.

Hummon, D.M. (1992). Community Attachment. In I. Altman and S.M. Low (Eds.), *Place Attachment*. Boston, MA: Springer, pp. 253–278.

Hwang, S.N., Lee, C., and Chen, H.J. (2005). The relationship among tourists' involvement, place attachment and interpretation satisfaction in Taiwan's national parks. *Tourism Management*, 26(2), 143–156.

Kasarda, J.D., and Janowitz, M. (1974). Community attachment in mass society. *American Sociological Review*, 39(3), 328–339.

Kerstetter, D.L., Confer, J.J., and Graefe, A.R. (2001). An exploration of the specialization concept within the context of heritage tourism. *Journal of Travel Research*, 39(3), 267–274.

Kozak, M. (2003). Measuring tourist satisfaction with multiple destination attributes. *Tourism Analysis*, 7(3/4), 229–40.

Krešić, D., Mikulić, J., and Miličević, K. (2013). The factor structure of tourist satisfaction at pilgrimage destinations: The case of Medjugorje. *International Journal of Tourism Research*, 15(5), 484–494.

Kyle, G., and Chick, G. (2002). The social nature of leisure involvement. *Journal of Leisure Research*, 34(4), 426–448.

Kyle, G.T., Absher, J.D., and Graefe, A.R. (2003a). The moderating role of place attachment on the relationship between attitudes toward fees and spending preferences. *Leisure Sciences*, 25(1), 33–50.

Kyle, G., Graefe, A., Manning, R., and Bacon, J. (2003b). An examination of the relationship between leisure activity involvement and place attachment among hikers along the Appalachian Trail. *Journal of Leisure Research*, 35(3), 249–273.

Lalli, M. (1992). Urban-related identity: Theory, measurement, and empirical findings. *Journal of Environmental Psychology*, 12(4), 285–303.

Lee, C.C. (2001). Predicting tourist attachment to destinations. *Annals of Tourism Research*, 28(1), 229–232.

Moore, R.L., and Graefe, A.R. (1994). Attachments to recreation settings: The case of rail-trail users. *Leisure Sciences*, 16(1), 17–31.

Moscardo, G. (1999). *Making Visitors Mindful: Principles for Creating Quality Sustainable Visitor Experiences through Effective Communication*. Illinois, IL: Sagamore Publishing.

Pieniak, Z., Verbeke, W., Vanhonacker, F., Guerrero, L., and Hersleth, M. (2009). Association between traditional food consumption and motives for food choice in six European countries. *Appetite*, 53(1), 101–108.

Pina, R., and Dias, Á. (2021). The influence of brand experiences on consumer-based brand equity. *Journal of Brand Management*, 28, 99–115.

Poço, T., and Casais, B. (2019). City branding and place attachment: A case study about Viana do Castelo. Proceedings of the *2nd International Conference on Tourism Research*. Porto, 14–15 March, pp. 226–233.

Proshansky, H.M. (1978). The city and self-identity. *Environment and Behavior*, 10(2), 147–169.

Ramkissoon, H., Smith, L.D.G., and Weiler, B. (2013). Testing the dimensionality of place attachment and its relationships with place satisfaction and pro-environmental behaviours: A structural equation modelling approach. *Tourism Management*, 36, 552–566.

Ribeiro, J.C., Vareiro, L., and Remoaldo, P.C.A. (2012). The host-tourist interaction in a world heritage site: The case of Guimarães. *China-USA Business Review*, 11(3), 283–297.

Santos, V., Dias, A., Ramos, P., Madeira, A., and Sousa, B. (2021). Mapping the wine visit experience for tourist excitement and cultural experience, *Annals of Leisure Research*, https://doi.org/10.1080/11745398.2021.2010225

Sarason, S.B. (1974). *The Psychological Sense of Community: Prospects for a Community Psychology*. San Francisco, CA: Jossey-Bass.

Scott, D., and Shafer, C.S. (2001). Recreational specialization: A critical look at the construct. *Journal of Leisure Research*, 33(3), 319–343.

Shams, S.R., and Lombardi, R. (2016). Socio-economic value co-creation and sports tourism: evidence from Tasmania. *World Review of Entrepreneurship, Management and Sustainable Development*, 12(2–3), 218–238.

Sousa, B., and Rocha, A.T. (2019). The role of attachment in public management and place marketing contexts: A case study applied to Vila de Montalegre (Portugal). *International Journal of Public Sector Performance Management*, 5(2), 189–205.

Sousa, B.B., Magalhães, F.C., and Soares, D.B. (2021). The role of relational marketing in specific contexts of tourism: A luxury hotel management perspective. In B.B. Sousa, F.C. Magalhães, and D.B. Soares (Eds.), *Building Consumer-Brand Relationship in Luxury Brand Management*. Hershey, PA: IGI Global, pp. 223–243.

Stokols, D., and Shumaker, S.A. (1981). People in places: A transactional view of settings. In J.H. Harvey (Ed.), *Cognition, Social Behaviour and the Environment*. Hillsdale, NJ: Lawrence Erlbaum Associationa, pp. 441–488.

Tonge, J., Ryan, M.M., Moore, S.A., and Beckley, L.E. (2015). The effect of place attachment on pro-environment behavioral intentions of visitors to coastal natural area tourist destinations. *Journal of Travel Research*, 54(6), 730–743.

Vareiro, L.M.D.C., Remoaldo, P.C., and Ribeiro, J.A.C. (2012). Residents' perceptions of tourism impacts in Guimarães (Portugal): A cluster analysis. *Current Issues in Tourism*, 16(6), 535–551.

Yoon, Y., and Uysal, M. (2005). An examination of the effects of motivation and satisfaction on destination loyalty: A structural model. *Tourism Management*, 26(1), 45–56.

Yüksel, A., Yüksel, F., and Bilim, Y. (2010). Destination attachment: Effects on customer satisfaction and cognitive, affective and conative loyalty. *Tourism Management*, 31(2), 274–284.

14 Reimagining Tourism Events

Spain's Preparation for the Return of a Healthier Breed of Tourists

Nuria Recuero-Virto

Introduction: Tourism Events Industry at a Tipping Point

Travel and tourism used to account for 12% of the gross domestic product (GDP, in advance) of Spain, which meant that Spain was ranked as the second destination in the world in terms of tourism arrivals and receipts (United Nations World Tourism Organization, 2021). Given such a significant role played by tourism activity in the country, understanding how the event industry has been affected by COVID-19 pandemic is a relevant aspect so as to acknowledge the present tourist behaviour and the future challenges (Gholipour et al., 2020). The effect of the management of the COVID-19 health crisis on the tourism industry is unprecedented mainly in terms of the volume of tourist arrivals, which has a great impact on the economies of all the destinations (Arbulú et al., 2021).

The year 2020 was the year that redefined the event industry. The COVID-19 pandemic has drastically changed our way of socializing and, thus, the tourism landscape. In the case of Spain, the state of alarm and emergency lasted for 99 days, which implied a strict lockdown where any kind of tourist movement was forbidden (inbound and outbound). As a consequence, the number of international tourist arrivals dropped to zero and the economic loss reached about 43.47 billions of euros (Sánchez-Pérez et al., 2021). Thus, it is a fact that the industry has been one of the most affected sectors. Event planners, who are professionals used to reinventing themselves, now have to confront two main trends that have come to stay, namely new safety concerns and the rise of virtual and technological hybrid events.

Recently, we have begun to see important face-to-face events coming back, such as Fitur and the World Mobile Congress (WMC), that are smartly dealing with the virtual dimension. This new hybrid panorama is offering the possibility of increasing presence of speakers and participants as many of them could not physically assist due to financial, conciliation or sanitary reasons. The risk of contagion has risen the popularity of events that entail less participants than before the pandemic, which is not only ensuring the social distance but also is permitting an improvement in the personalization and the human connection during the events. As a result, partnerships and hence stakeholder collaboration have become more critical than ever to recover and achieve a successful hybrid event.

DOI: 10.4324/9781003271147-19

Due to uncertainty, there is a strong need to work jointly to forecast and act for possible scenarios. The last lockdowns and first face-to-face events have been a process for events planners of intensive learning, which have brought out the identification of some problems that need to be solved, such as videoconferencing saturation, COVID tracking, among others. In contrast, other kind of initiatives have arisen. For instance, Ibiza has recently diversified its portfolio of tourism products by adding to nightlife tourism activities, sports tourism, in particular mountain bikers' races (Rejón-Guardia et al., 2020).

Innovativeness is vital for creating expectation and surprising and unique experiences, which result in unforgettable events. Therefore, the linkages and connections the events create have to be authentic, now more than ever due to the unprecedent situation we have suffered. The digital needs of tourists have increased and they are now expecting creativity, a hybrid social engagement. Tourists are demanding mobile attractive content that ensures their contactless sanitary secureness and that entertains them. Thus, in this chapter the future tourism event trends will be described, which are connected to the new desires and needs of this new generation known as Generation Clean.

Spanish Post-COVID-19 Patterns

Despite the fact that the uncertainty is great, the forecast for the upcoming years reflects a notable increase in the number of events (as many where postponed), but a decrease of participants and spending (Vuletin, 2020). This disruptive pandemic has had an unexpected impact on tourism flows, behaviour, expenditure, travel choice, perceived risk and the redefining of tourist satisfaction (Sánchez-Pérez et al., 2021). This has meant the appearance of new tourism patterns that have come to stay. The most significant changes will be explained below.

Generation Clean

Due to the "new normal", postponements and cancellations of in-person events may still take place. Until effective vaccines are available, the reputation of the brands, and indeed of event organizers, will be built on health and safety. New safety measures will still be adopted so as to fulfil the needs of Generation Clean. A duty-of-care strategy will remain at the forefront from policy design to business planning as onsite protocols will be critical factors for face-to-face events, and safety measures are paramount.

Thus, the next actions have been adopted as during live events there is still a substantial risk of contracting COVID-19:

• social distancing is the preference for open air-spaces;
• monitoring on-site testing or request of updated COVID-19 related information in QR codes (vaccine, PCR test or others) for attendees and employees;
• temperature check support for attendees and employees;
• mask wearing compulsory indoors and handwashing or disinfecting stations;

- allocated seating spaces and separated spaces by plexiglass or other materials, especially for meal areas and preferably food will be skipped to avoid unnecessary risks;
- stewards to control attendees and foster obedience of safety measures;
- cleaning frequency planned;
- real-time COVID-19 related data;
- Bring Your Own Device (BYOD) initiatives to ensure contactless is a priority (such as QR codes with COVID-19 related information or facial recognition);
- quality stamps that ensure that safety protocols have been implemented;
- events bubbles are also guaranteed by a technological system that ensures constant tracking of COVID-19 information.

Tourists will ask for destinations that are secure in sanitary terms and that are medically prepared and within almost all the population vaccinated. Thus, Spain reunites all these requirements.

Micro Experiences

Experiences that cater for a small number of persons will be considered less risky, more personalized, exclusive and intimate than large events as these imply a lot of complications. This will mean an opportunity to redefine in person connections during events so these can be meaningful. Events will be measured taking into account the number of connections they enable. Thus, this context will increase competition and will push the limits of creativity to foster engagement. All this context is not only highly segmenting audiences but also is resulting in a strong tendency towards personalization. Thus, tailor-made experiences happen to be in vogue and now destinations that offer small spaces for these kinds of events are demanded by planners. Moreover, micro experiences offer the possibility to surprise and excite attendees as ever before but as a disadvantage the prices for tourists will inevitably rise.

Several event organizers have realized the opportunity of creating a niche-market of events for very specific topics, which build a community towards those themes. These smaller events with fewer participants could be more frequent and avoid the peaks of tourism flows (or even tourism seasonality) and increase interaction. The rise of frequency will result in higher rates of engagement as event planners will be able to personalize and hence attendees will perceive experiences as unique and memorable. In this regard, Barcelona is achieving this objective by focusing tourism economic recovery on the reactivation of corporate travel business and locating unique and singular spaces for these sorts of events. These spaces have evolved into innovative areas equipped with the most advanced technology and centred on sustainable purposes, which is emphasizing Barcelona's role in the tourism event industry.

We are all facing an aging world, where the elderly has been the segment most affected because of the pandemic not only in terms of losses but also regarding the willingness to travel. As part of the new normality, tourists are demanding destinations that can organize open-air activities (and thus, with a favourable climate) and less mass tourism. Spain is in a position to take advantage of these new requirements (Sánchez-Pérez et al., 2021).

Hybrid Tech Events

As a way to ensure attendance, events will continue to have a mixed format as online audience is always guaranteed. In-person events will be more segmented and smaller than before the pandemic and the budget per event will unfortunately tend to decrease, due to the economic crisis, as the number of events increase but with fewer participants. Despite the limited budget, the format of these events is now transforming into a TV show that relies on live speakers, shorter sessions and a strong tech infrastructure that increases the production quality. Precisely, new tech devices have replaced other priorities such a square metre space to support on-site and virtual connections such as improving VR (i.e., virtual reality), AR (i.e., augmented reality), video meetings, chats, polls access, among others. The budget spent before in food and beverage will be dedicated to the investment for digital transformation and for broadcasting the hybrid event production.

The challenge is winning the race to develop and implement the virtual event platform and thus now many partnerships are taking place. Deeper understanding of attendees' changing behaviour is required, and hence the success of events will depend on the ability to adapt to the events by using technological advancements such as AI (i.e., artificial intelligence), 360° videos, among others, that avoid screen fatigue and entertain. The reduction of budgets and the increased expectations as the audiences are increasingly demanding available technology and quality shows rather than simple events is complicating the labour of event planners. In addition, the duality of the events (online and offline) is making the logistics more sophisticated, which is intensifying planning and strategy that takes into consideration possible technology failure and unexpected emergencies. Besides, these events are now entailing two audiences: the online users and the in-person attendees, that have different needs to be covered.

Virtuality

The virtual dimension of events is yet a challenge for organizers as it is planned at the same time live in-person experience takes place. Event organizers have to prepare live and on-demand content for its clients, which raises up the level of sophistication as it entails the preparation of two experiences that require fulfilling the needs of each audience. Attendees' needs are getting more sophisticated. They are demanding synchronous (i.e., live-streaming) and asynchronous content in platforms that have an easy-to-use interface, that display three or four speakers' views on livestreaming, overlays, popups content, among others. The trend is moving from webinars, live streaming and content creation, which lacked human interaction, to events that entail on-stage representation timely structured to foster connections among attendees. Thus, the metrics of engagement will be measured regarding the connections made during events. Social media channels will be more interactive and will improve the two-way communication. The expectations are rising. The market is growing.

Bundling could be an option to clients that demand an added value service and it could be performed by even generating a subscription model, better content on demand or one-to-one meetings that strengthen connection possibilities for business objectives. Corporate clients may prefer virtual events as it saves money and reduces the time to reach the destination, for just a two-hour meeting. Moreover, the virtual dimension reduces the environmental impact, as it has been proved with the rapid drop of carbon footprint, as well as avoiding the threat of contracting COVID-19.

Dealing with Fear of Missing Out (FOMO) in Favour of Purposeful Events

The current climate of uncertainty is leading to prioritize purposeful events worldwide and particularly in Europe that is facing an unprecedent scenario due to Brexit, political instability and weather emergencies. Specifically, in the event industry, Spain is mainly confronting the consequences from the economic slowdown, the volcano's eruption in La Palma, and the cancellation of the World Mobile Congress during the coronavirus crisis in 2020. Thus, in this unusual moment sustainability, specifically local social and environmental commitment, has gained popularity among events and local residents are more involved than before. After the isolation, everyone is willing to attend awesome sensorial purposeful events. The experience design will require that socializing not only the sanitary measures abovementioned, but also the possibility of creating community across connecting with other attendees rather than networking.

Sustainable and Diversity Priorities

In this new normality, there are two key priorities for the Spanish event organizers for the next years: fostering sustainability and diversity. The reduction of environmental impacts has been emphasized during the COVID-19 lockdowns and the diminution of the carbon footprint is now a policy to accomplish in every single worldwide company. A plethora of Corporate Social Responsibility (CSR) initiatives have been endorsed by the tourism event industry that deal not only with hygiene and safety protocols, but also with the meanings of the events, the topics and even the messages offered during them.

In this regard, Vuletin (2020) conducted a survey with a sample of more than 850 Spanish organizers of business, networking, musical, community, gastronomic, scientific and gala events, where it was pointed out that almost 70% of Spanish event organizers have made progress on environmental issues, although most are in the early stages of applying sustainability to their events. Specifically, it was found out that 84% use digital tickets and 50% employ biodegradable or reusable elements for food and beverages.

Creativity formats that entail contactless activities will be highly demanded (such as design BYOD, Bring Your Own Device; virtual design thinking sessions; hackathon videoconference meetings, among others) as ways to foster co-creation and collaboration. Event professionals will try to arrange these sponsorships with local businesses as the quarantine precautions have emphasized the need of ensuring local economies, as these are essential to guarantee sustainable development.

Concerning diversity, 50% are attracting assistants of different religions, ages, nationalities, among other aspects, and four out ten are looking for speakers that meet the interests of a diverse audience. In this respect, many event organizations have been recently establishing policies that provide the necessary education to address inclusion and diversity and also to ensure that event employees become mindful so as to improve attendee experience taking into account individual variances. Thus, many specific activities have been introduced such as the consideration of dietary particularities due to health or religious reasons, the representation of diverse speakers, among others. In other words, attendees' wellbeing will have maximum relevance and for all businesses the approach will be customer centred, with their satisfaction the most important priority towards guaranteeing sustainable tourism development.

Business Travel

It is undeniable that MICE tourism (Meetings, Incentives, Conventions/ Conferences and Exhibitions) is an essential pillar of the event industry as it encompasses the various types of group business travel (Rojas et al., 2020), a segment which is characterized in Spain by the following factors: high average expenditure, slight seasonal fluctuation, demanding qualified travellers, excellence of service quality, tailored service, predisposed to high-tech advancements (Rojas et al., 2020). Local economies benefit from corporate travel as these tourists normally have a big budget to spend on the destination. However, the consequences due to the pandemic for many companies are resulting in a need of cutting luxuries such as business travel.

This segment is particularly interesting for Spain, as it constitutes 15% of our tourism sector – which, overall, accounts for 12.4% of GDP – and generates more than 80,000 jobs. According to the Events Industry Council (2018), congress and event tourism generated a direct impact of more than ten billion euros on the Spanish economy in 2017, ranking us as the 12th country with the highest direct income generated due to MICE tourism. Before the pandemic, in 2019 this sector was growing: the number of meetings was 29,603, 14.7% more than the previous year; and meeting attendees reached 4.8 million, 11.4% higher than in 2018 (Events Industry Council, 2018). The two main sectors of activity are still the medical-health and economic-commercial industry; the months in which a larger meeting is held continue to be spring and fall and the annual average of stay is 2.16 days; and the percentage of men is still higher than that of women aged 35 to 54 years. Finally, *bleisure* (i.e., business and leisure) has risen in the last decade, where business events are combined with other tourism activities. Therefore, business travel is expected to grow in the upcoming years.

Conclusion: Lessons Learnt

COVID-19 has defined the reboot of the tourism activity since this industry has been one of the sectors most touched, and Spain one of the countries most affected (Sánchez-Pérez et al., 2021). This pandemic has seriously affected society's quality of life and the well-being of people, and thus it has implied changes in the tourism market. This chapter has pinpointed the future patterns the tourism event industry will have to deal with so as to be successful. To sum up, Spain has done its homework. During these first initial trials, we have seen that the country is medically ready to receive tourists, is positing as a destination for open-air micro personalized events, is prepared for amazing hybrid high-tech business travel events such as the WMC, and is mindful about rising awareness of the relevance of sustainability and diversity organizing purposeful events. However, the next challenge lies ahead, which is coping with the new technological advances so as to compete and generate online and offline engagement during the tourism events.

References

Arbulú, I., Razumova, M., Rey-Maquieira, J., and Sastre, F. (2021). Can domestic tourism relieve the COVID-19 tourist industry crisis? The case of Spain. *Journal of Destination Marketing & Management*, 20, 100568.

Events Industry Council (2018). *Global economic significance of business events study.* Retrieved from: https://insights.eventscouncil.org/Portals/0/OE-EIC%20Global%20Meetings%20Significance%20%28FINAL%29%202018-11-09-2018.pdf (accessed 23 October 2021).

Gholipour, H.F., Arjomandi, A., Marsiglio, S., and Foroughi, B. (2020). Is outstanding performance in sport events a driver of tourism? *Journal of Destination Marketing & Management*, 18, 100507.

Rejón-Guardia, F., Alemany-Hormaeche, M., and García-Sastre, M.A. (2020). Ibiza dances to the rhythm of pedals: The motivations of mountain biking tourists competing in sporting events. *Tourism Management Perspectives*, 36, 100750.

Rojas, A., Alarcón, P., and del Alcázar, B. (2020). The MICE tourism value chain: Proposal of a conceptual framework and analysis of disintermediation. *Journal of Convention & Event Tourism*, 21(3), 177–200.

Sánchez-Pérez, M., Terán-Yépez, E., Marín-Carrillo, M.B., Marín-Carrillo, G.M., and Illescas-Manzano, M.D. (2021). The impact of the COVID-19 health crisis on tourist evaluation and behavioural intentions in Spain: Implications for market segmentation analysis. *Current Issues in Tourism*, 24(7), 919–933.

United Nations World Tourism Organization (UNWTO) (2021). *World Tourism Organization underscores tourism's importance for covid-19 recovery in audience with the king of Spain.* Retrieved from: www.unwto.org/news/unwto-underscores-tourisms-importance-for-covid-19-recovery-in-meeting-with-the-king-of-spain   (accessed   2 November 2021).

Vuletin, M. (2020). Informe tendencias de los eventos en 2020 by Eventbrite. *MICE.ES.* Retrieved from: www.mice.es/post/789-informe-tendencias-de-los-eventos-en-2020-by-eventbrite (accessed 27 October 2021).

Part V
Cases from North America

15 Surviving the COVID-19 Pandemic

How Technology is Getting the Tourism Industry Back on Its Feet in the USA

Elanie Steyn and Imran Hasnat

Introduction

January 2020 saw the start of a series of events that affected the world in ways that few could have predicted. When the World Health Organization announced on 11 March 2020 that "COVID-19 can be characterized as a pandemic", there were already "more than 118,000 cases in 115 countries, and 4,291 people have lost their lives". Several thousands more were "fighting for their lives in hospitals" (World Health Organization, 2020).

The months that followed saw countries around the world having to face economic, health, social and human rights challenges that left many struggling. A Congressional Research Service Report (2021) indicates that the virus "reduced global economic growth in 2020 to an annualized rate of -3.4% to -7.6%". Even though some forecasts showed that growth rates would increase toward the end of 2020, the lingering nature of the pandemic and its continued impact on all sectors of society made these predictions hard to achieve.

One of the industries hit the hardest by the pandemic is the tourism industry (United Nations World Tourism Organization (UNWTO), 2021). According to the UNWTO, before the pandemic the tourism industry was the third-largest export category in the world's economic sectors; it contributed more than 10% of global GDP and added more than 320 million jobs worldwide in 2019. In some countries, these numbers could be even higher. However, the UNWTO (2021) predicts that "export revenues from tourism could fall by $910 billion in 2020, … and reduce global GDP by 1.5%".

The United States is one of the most popular tourism destinations in the world. As stated in a 2021 Statista Report, almost 80 million tourists visited the USA in 2019, helping the travel and tourism industry to contribute over US$1.1 trillion to the country's GDP. However, the pandemic resulted in total travel spending in 2020 declining by 42% from the previous year. While this decline was significant for international travel spending, domestic travel also suffered (US Travel Association, 2021). These declines did not only affect the economy, but also impacted job creation, tax revenue and related sources of income.

DOI: 10.4324/9781003271147-21

The relationship between technology and tourism is not a new one (Neuhofer et al., 2013). In recent decades, though, with the introduction and increasing popularity of Information and Communication Technologies (ICTs), technology has revolutionized the way travel is planned, business is conducted and tourism services and experiences are created and consumed (Neuhofer et al., 2013). The period described above and the impact of the pandemic on the tourism industry has propelled the need for the tourism industry to use technological solutions in a faster and more innovative way. This reality has presented the industry with both opportunities and challenges related to the use of technology-related solutions.

This article outlines the role of technology in helping the US tourism industry recover and innovate as the Covid-19 pandemic continues. It shows how technology, applications and big data are helping tourism operators get back on their feet and encourage travellers to explore with confidence amidst a global crisis that has brought many industries to their knees.

The Role of Tourism Globally

The tourism industry is without doubt one of the most significant players contributing to growth and development globally. Not only does it contribute to GDP, jobs and economic growth in general, but it has also been a significant player in changing perceptions about countries and peoples, preserving local cultures and empowering indigenous peoples and communities (Hasnat and Steyn, 2021). It provides opportunities for foreigners to learn about the people and the culture they visit but also provides avenues for local businesses and entrepreneurs to sell products and services to a market that is more sustainable than just selling to local residents (Yehia, 2019).

According to the World Travel and Tourism Council (2021), in 2019, travel and tourism contributed to 25% of all new jobs created across the world, almost 11% of all jobs and close to 11% of global GDP. This impact is even more significant in some countries, especially those in the developing world.

For the Maldives, for instance, the millions of people visiting the country annually contribute to more than a third of the country's GDP, making it the country in the world most reliant on tourism expenditure (Statista, 2020; Frost, 2019). Similarly, in Bangladesh, it is estimated that nine jobs are being created for every one tourist that visits the country (Madden, 2020). And, in India, even at the beginning of the twenty-first century, the travel and tourism industry created more jobs than automotive manufacturing, chemical manufacturing, communications and the mining sectors combined (see Simpson, 2012). By 2020, jobs created by the tourism industry in India reached 39 million or 8% of the total employment in the country (India Brand Equity Foundation, 2021) and India saw the strongest growth in the number of jobs created through tourism, followed by China and the Philippines. It is estimated that every tourist to India creates the opportunity for at least two jobs (Madden, 2020).

However, as the COVID-19 pandemic continues to affect industries worldwide, the tourism industry is, as could be expected, one of the hardest hit. While global GDP is expected to fall by 1.5% due to the pandemic, export revenues from global tourism is said to fall by $910 billion in 2020 (United Nations World Tourism Organization, 2021), with an estimated 50 million jobs at stake globally (Lau, 2020). As this organization also points out, some countries (especially smaller, developing countries) will be affected most, as tourism traditionally accounts for as much as 80% of exports in those countries. Similarly, it is estimated that as many as 100 million direct jobs could be lost in tourism as a result, with industries related to tourism (e.g., accommodation and food services) and the small business industries also being severely affected.

Using the same three countries above as examples, the Maldives, for instance, saw a drop of almost 68% in tourism in 2020, forcing authorities to think innovatively as it opened its borders after several months. As such, the government is implementing efforts to diversify the country's economy and focus on industries such as fisheries and agriculture so as to not rely on tourism as heavily as it did in the past (Choudhury, 2021).

In Bangladesh, the Ministry of Civil Aviation and Tourism indicated that, in addition to the significant financial losses the country's tourism industry has suffered since the start of the pandemic, more than 4 million tourism-related jobs had been lost and about 70% of people associated with the industry face an uncertain future (Islam et al., 2021). As Polas, Saha and Tabash (2021) point out, hotel occupancy in Bangladesh had dropped from 80% in pre-pandemic times to 30% since the pandemic. In addition, while tourism-related activities had been completely stopped for several months, business owners still have to pay rent, utilities and compensate core human resources. In addition, a study among more than 300 potential tourists to Bangladesh has found that the pandemic has an additional impact on tourism in that tourists' knowledge and perceived health risk when traveling to the country could increase their hesitation to travel. This has implications for how tour operators, hotel managers and staff and marketers should approach tourists and potential tourists to the country (Polas et al., 2021).

Jaipuria, Parida and Ray (2021) point out that in March 2020, India saw a decline of almost 67% in tourist arrivals compared to the same time in 2019. They state that, as a result, India is likely to lose around 40 million jobs and about $17 billion in revenue annually because of the pandemic. This reality had led FAITH (the Federation of Travel and Hospitality Businesses in the country) to petition the government for several waivers to tourism and hospitality businesses for utility costs, property taxes and automatic renewal of licenses, permits and permissions without financial charges (Mathur, 2021). Despite these accommodations, many in the sector still struggle because of a lack of cash flow. Similar to strategies adopted in Bangladesh, role-players in the Indian tourism industry are working toward implementing tactics that would attract tourists to India once borders are fully opening again. These include informational campaigns through media and social media outlets, encouraging tourism operators to be flexible in their booking and cancellation policies, and attracting tourists to remote areas of the country that are less populated but still offer a variety of tourism activities and scenic beauty (Shetty, 2021).

The Role of Tourism in the United States

The United States is one of the most popular tourism destinations in the world. The millions of tourists who visit the country every year contribute to just under $2 billion in economic output, which supports close to 8 million jobs or 1 in 10 jobs in 2019 (see Statista, 2021; SelectUSA, 2021; World Travel and Tourism Council, 2021). In 2019, 26% of inbound arrivals were from Canada, followed by Mexico with 23% (World Travel and Tourism Council, 2021).

However, the country is not only popular among foreigners. According to the US Travel Association (2020), US domestic travel had increased "1.7% from 2018 to a total of 2.3 billion person-trips in 2019". Almost 80% of those trips were focused on leisure. Not only does tourism have a direct economic impact on jobs, payroll income and tax revenues, it also provides several opportunities for local entrepreneurs.

According to WorldAtlas (2018), the five most popular states for tourists in the USA are California, Florida, Nevada, Texas and New York because of their cities, entertainment, natural attractions, architecture, museums and more. When it comes to domestic travel, a survey by HotelsCombined, a hotel booking site, that analysed more than 87,000 hotel bookings that originated in the USA, found that "New Yorkers book the most hotel rooms in Florida, Texans like to travel to California, and Californians like to travel to Nevada … but no state wants to visit Alaska" (Polland, 2014).

As is the case in other parts of the world, the pandemic has affected travel in and to the USA significantly. According to the US Travel Association (2021), total travel spending in the USA in 2020 declined by 42% from the previous year, with international travel spending falling by 76% and domestic travel spending declining by 35%. Similarly, business travel spending in the USA saw a 70% decline compared to 2019. As the US Travel Association (2021) points out, between March and December 2020, the pandemic has been responsible for "$492 billion in cumulative losses for the US travel economy". This decline was seen across the board, almost, with 18 states and territories experiencing a downturn of more than 40% in travel spending (US Travel Association, 2021). As could be expected, popular tourist destinations such as Hawaii and Nevada suffered the most, seeing a 60% downturn in travel spending and losses of thousands of jobs as a result of the pandemic.

The above impact has, as Henderson (2020) points out, created "economic crises" in some states most dependent on tourism. Not only do these states see less tourism spending, but they also suffer a cumulative effect as hotel rooms and convention centres were left empty, events were cancelled and overall tax revenue declined.

Amidst these changes and widespread impact, the Organisation for Economic Cooperation and Development (OECD) (2020) states "the crisis is an opportunity to rethink tourism for the future. Tourism is at a crossroads and the measures put in place today will shape the tourism of tomorrow".

One key to survival for role-players in the tourism industry seems to be flexibility and the ability to think innovatively about how they provide products and services to their customers (Henderson, 2020). Technological advances are playing a key role in this process (see Sharma et al., 2021) and will most likely continue to do so into the future.

The Role of Technology in the Tourism Industry Before and During the COVID-19 Pandemic

As long as a decade ago, writers and scholars have pointed out that the hospitality industry faces "a quintessential question" of capitalizing on the advantages that technology provide the industry (in the ability to replace human resources with less expensive technology) while balancing the challenges of technology. These challenges focused on technology replacing human contact, having an impact on the environment or becoming an easy way to breach security and promote terrorist activities (Brain, 2011; Hojeghan and Esfangareh, 2011; Kim and Kim, 2017; Magio, 2018).

In spite of these challenges, the reality is that travel and technology have become an inseparable combination in recent years. A 2019 Google Travel study indicated, for instance, that as many as 74% of travellers had moved from using travel agencies to plan their travel to using the Internet to do so (Vidal, 2019). As Vidal (2019) highlights, millennials' combined passion for travel and interest in technology, as well as the travel industry capitalizing on the benefits of technology across a variety of fronts, have further contributed to this relationship becoming more integrated.

Even before the start of the COVID-19 pandemic, leaders in the tourism industry worldwide have indicated that the interrelationship between technology and the industry was bringing about a transformation "of the tourism system itself" (Vidal, 2019) as it responds to the needs and concerns of a "traveller more concerned with sustainability, and with more technology or curiosity" (Vidal, 2019). However, the wide range of disasters related to the pandemic has significantly sped up this transformation and has made the industry's dependence on technology even more critical. As Sharma, Thomas and Paul (2021) point out, technology such as artificial intelligence, Big Data analytics and applications on mobile phones, as well as more powerful data networks, faster download speeds, wider network coverage and more stable connections, played a role not only to make the industry more flexible but also to reduce cost (Vidal, 2019).

When COVID-19 shut down most of the travel and tourism industry worldwide from January 2020, role-players in these sectors had to think innovatively and adapt their business models and product offerings. This was not only necessary due to the immediate and tangible impact of the pandemic on the industry, but also because a study showed that only 9% of Americans ranked the industry as the "most innovative" during the pandemic (see Fox, 2020). Much of the innovative thinking that took place as a result was made possible because some businesses capitalized on the benefits that new technology presented to them, putting the concerns related to privacy and dependency on the backburner.

Companies across the world adjusted their products and services by thinking differently about information, system and service quality, how they themselves (and their customers) use technology to their advantage, how they measure both customer satisfaction and the benefits technology presented to the industry as a whole (see Lau, 2020). In addition, some organizations started to utilize technology more as they worked to increase (and in some cases improve) brand awareness, reduce safety risks and build relationships, as industry experts point out that customers' expectations of good experiences would be higher than ever after returning to a "new normal" (Fox, 2020).

As Lau (2020) indicates, investing in new technology with the view of surviving the challenges the pandemic presents and having a competitive edge in the long term, might give the tourism and hospitality industry the advantage of embracing equipment and systems that improve their business models as well as their inter-action with different constituencies. "Technology features originally intended as novelties are becoming necessities during times when some people are wary of even stepping outside their homes" (The Economic Times, 2021). A study by Sitel Group (Fox, 2020) showed consumers believe digital representatives and virtual reality were two of the leading technologies the industry can use to improve customer experiences.

Examples of these and other initiatives are discussed below.

Live streaming, Augmented Reality and Virtual Reality

In the same way that educational and other institutions have utilized these cap-abilities to reach customers and audiences, the hotel industry capitalized on these online trends to attract customers and promote their products and services. The World Economic Forum (2021) indicated that the pandemic might be the event that has prepared the world for "virtual tourism" and that this concept could be a significant role-player as the world prepares to return to travel (albeit in a different way) both during and after the pandemic. Augmented, virtual or mixed reality (AR, VR, MR) could play a significant role in this process (see Haugen, 2020) and the World Travel and Tourism Council (2021) has, in fact, asked governments to increase their budgets to promote their countries as tourism destinations. Virtual Reality plays an important role in this process (see Rogers, 2020; Akhtar et al., 2021). As Lau (2020) points out, celebrities give tours of facilities (guest rooms, banquet halls, conference facilities) to attract customers and show what establishments can offer them, even though they cannot visit the facilities in-person for tours. In China, for instance, companies utilized the tech-nology, facilities and experience that big hotels have in hosting virtual conferences, keeping attendees safe, saving time and money that virtual attendance brings while still giving opportunities for networking and exchange of ideas.

Similarly, with the world in lockdown, companies with Virtual Reality booking systems saw a dramatic increase in the amount of VR downloads during the pan-demic. As such, some downloads increased by 60% between December 2019 and April 2020, and doubled again between January and April 2020 (Pierce, 2021).

These downloads enabled visitors to visit national parks, museums, international cities and landmarks, and even multi-day trips to foreign locations.

While these applications and experiences kept customers busy, entertained and informed at a time when they could not physically travel, it also created the opportunity and interest for potential "real" visits when the pandemic becomes more manageable and people are able to travel again. Moreover, these tools could play an important role in country and destination marketing efforts and provide employment for "local tourist guides, artisans and others as global citizens in the tourism industry" (World Economic Forum, 2021).

As Debusmann (2020) puts it: "digital applications cannot, and are not intended to, replace the experience of real-world travel … but VR and AR applications are essential elements in keeping interest (in a destination) alive … getting potential customers excited about our product and providing inspiration for real-world travel".

Software applications

Several role-players in the tourism and hospitality industry are capitalizing on the benefits of software applications to address the challenges COVID-19 presents (Baratti, 2021). While some organizations are developing new software applications to solve these problems, others, such as the airline KLM, started using existing applications such as Facebook Messenger to communicate with passengers. This communication ranges from sending boarding passes, messages and status updates to passengers, enabling them to access information on their personal devices (Vidal, 2019). Applications like these alleviate the pressure on airline call centres and give travellers more control over their scheduling, rebooking and refunding processes if necessary (see PlugandPlay, 2020). In Summer of 2020, Google Assistant launched an application specifically aimed at the hospitality industry and focusing on voice-command applications in hotel rooms. This application was introduced in high-end hotels in New York City and Scottsdale, Arizona (Zambello, 2021).

Similarly, when restaurants had to close for dine-in services, many diversified and started focusing on food delivery or sales of bulk meal boxes to customers using software applications that either have been in use already or gained more popularity because of COVID restrictions.

Cloud-based technology can further help hotels and restaurants coordinate housekeeping systems, staffing assignments and compliance with standards for hygiene and cleanliness (The Economic Times, 2021).

Data and Augmented Reality

The hospitality and tourism industry is implementing data and machine learning in a variety of ways, ranging from helping them predict demand and occupancy rates, and better manage revenue models. In addition to this technology addressing the need for staff to play this role (in an environment where human interaction had been limited for several months), it also helps in that staff who did focus on these roles can now adopt a more strategic thinking and planning role (see Pierce, 2021).

Big data also allows agencies to develop automated travel agent networks, to a much bigger extent than platforms such as Booking.com had done in the past. These capabilities allow for customers to have a personalized and autonomous agent (through AI) who search on their behalf and based on their personalized preferences (previous bookings, rates paid, etc.) (Dotson, 2021).

Players in the hotel and restaurant industry in China, for instance, are taking advantage of the benefits data tracking and AI present them to do facial recognition when guests check in at their establishments. These abilities allow them to monitor temperature, use facial-recognition cameras and communicate with authorities that request or require data on the health status of visitors (see Lau, 2020; also see Baratti, 2021; Gössling, 2020).

Digitalization and contactless activity

Though much of customers' experience in the hospitality industry has always been based on face-to-face and human interaction (see Gössling, 2020), a lack of human contact and interaction had been a key element of efforts to contain the pandemic (Zambello, 2021). As a result, hotels across the world, including some in Colorado and Idaho in the USA, had introduced "contactless, express check-in" through digital key access (Hospitality Technology, 2021). Similarly, a hotel in New Orleans introduced the services of "robot butlers" who deliver newspapers, drinks and snacks to guests' rooms, reducing face-to-face interaction even further. In addition, as Hospitality Technology (2021) points out, this service has become a source of additional revenue for the hotel, as it charges a delivery fee for some products. However, Pierce (2021) points out that many of these "robot butler" ideas have not been successful yet.

Voice assistant solutions also enable guests to adjust lights or the thermostat in their rooms and control in-room entertainment without having to touch controls or switches (Berger, 2020).

Cleaning technologies

Hotels and other hospitality providers rely heavily on staff for cleaning rooms and other common areas. Given restrictions related to contact and interaction, some hotels have implemented cleaning technologies that would disinfect spaces and other technologies that use UV light to hone in on bacteria and help prevent these from spreading (see Zambello, 2021; Castellanos, 2020).

Conclusion

The COVID-19 pandemic has changed the face of several industries and will most likely continue to do so in the foreseeable future. The tourism and hospitality industry is no exception and probably one of the industries hit hardest by the pandemic. Its very nature depends on human interaction and exchange – something the pandemic made impossible. Though this industry was in initial phases of adopting technology to play a role in its transformation and keeping up with consumer needs and wants, the pandemic left it with no choice but to think innovatively and fast. As Debusmann (2020) states, "the pandemic has been 'a shot of adrenaline'" for the adoption of new technology in this industry. Instead of using technology to make travel more convenient in the pre-Covid era, the industry is now looking toward technology to make travel "safer, more manageable, or even possible" (Digiteum, 2021).

This chapter outlined the economic and other roles of the tourism and hospitality industry worldwide, as well as in the USA. It also showed the impact the pandemic has on every level of this industry. Scholars and industry experts agree that technology is probably one of the most significant ways in which the industry can adapt and (potentially even) become more successful than it had been in the past. As Pierce (2021) points out, new technologies will be key post-pandemic for the hospitality industry to manage revenue in an automated way, target niche audiences in a focused way, forecast fares, prices and seasonal demands, personalize searches and bookings based on customers' preferences and personal history, and, probably most importantly, keep customers safe and healthy through limited human interaction and advanced cleaning technologies (also see Dorsi, 2021). It is predicted that several of the technologies originally developed as "emergency" measures when the pandemic hit, will probably become a part of the travel and tourism industry of the future. These include contactless services at airports, hotels and so on; digital identification through health and vaccine passports, for instance; more secure and flexible travel planning and management technologies; robots that assist with different aspects of the travel experience; and the Internet of Things (IoT) and big data to make travelling more secure and help with tracking and predicting of events (see Digiteum, 2021).

According to Lio Chen, Senior Vice President for US-based platform PlugandPlay, to successfully implement technology in reviving the travel and tourism industry, a key factor is that it should "work together on open data initiatives … airlines, airports and hotels must break all the data silos and work together with more companies" (PlugandPlay, 2020). These types of collaborations will not only help the travel and tourism industry get back on its feet but will most likely present the industry with technological solutions that will benefit it for a long time to come.

References

Akhtar, N., Khan, N., Khan, M.M., Ashraf, S., Hashmi, M.S., Khan, M.M., et al. (2021). Post-COVID-19 tourism: Will digital tourism replace mass tourism? *Sustainability*, 13, 5352.

Baratti, L. (2021). Technology's impact on the future of travel post COVID-19. *TravelPULSE*. Retrieved from: www.travelpulse.com/news/travel-technology/technologys-impact-on-the-future-of-travel-post-covid-19.html (accessed 12 August 2021).

Berger, D. (2020). Innovative ideas for keeping guests safe during the pandemic. *hospitalitynet*. Retrieved from: www.hospitalitynet.org/opinion/4100555.html (accessed August 2021).

Brain, L. (2011). Tourism in the world of technology. *TourismReviewNEWS*. Retrieved from: www.tourism-review.com/the-world-of-technology-affects-tourism-industry-immensely-news2757 (accessed 25 August 2021).

Castellanos, S. (2020). Hospitality industry turns to tech to lure guests back. *The Wall Street Journal*. Retrieved from: www.wsj.com/articles/hospitality-industry-turns-to-tech-to-lure-guests-back-11596636001#:~:text=-based%20Viceroy%20Hotel%20Group,%20said,minuscule%20droplets%20adhere%20to%20surfaces (accessed 20 August 2021).

Choudhury, S.R. (2021). As tourism picks up again, Maldives steps up efforts to diversify its faltering economy. *CNBC*. Retrieved from: www.cnbc.com/2021/03/02/tourism-dependent-maldives-steps-up-economic-diversification-effort.html (accessed 24 August 2021).

Congressional Research Service (2021). *Global economic effects of COVID-19*. Retrieved from: https://sgp.fas.org/crs/row/R46270.pdf (accessed 20 August 2021).

Debusmann, B. (2020). Coronavirus: Is virtual reality tourism about to take off? *BBC News*. Retrieved from: www.bbc.com/news/business-54658147 (accessed 20 August 2021).

Digiteum (2021). *5 Technologies for travel and tourism industry in post-COVID era*. Retrieved from: www.digiteum.com/technologies-travel-tourism/ (accessed 19 August 2021).

Dorsi, S. (2021). Technologies in travel help revive the industry. *TourismReviewNEWS*. Retrieved from: www.tourism-review.com/technologies-in-travel-are-essential-for-boosting-the-industry-news12052 (accessed 11 August 2021).

Dotson, K. (2021). Fetch.ai is bringing AI travel agents to 770,000 hotels worldwide. *SiliconANGLE*. Retrieved from: https://siliconangle.com/2021/01/15/fetch-ai-bringing-ai-travel-agents-770000-hotels-worldwide/ (accessed 20 August 2021).

The Economic Times (2021). *How hotels are using technology to boost traveller confidence during COVID times*. Retrieved from: https://economictimes.indiatimes.com/industry/services/hotels-/-restaurants/how-hotels-are-using-technology-to-boost-traveller-confidence-during-covid-times/articleshow/80906994.cms?from=mdr (accessed 18 August 2021).

Fox, J.T. (2020). Pandemic offers hotels opportunities for innovation. *HOTEL MANAGEMENT*. Retrieved from: www.hotelmanagement.net/tech/pandemic-offers-innovation-opportunities-for-hotels (accessed 28 August 2021).

Frost, N. (2019). These are the countries most reliant on your tourism dollars. The 25 most tourism-dependent countries. *Quartz*. Retrieved from: https://qz.com/1724042/the-countries-most-reliant-on-tourism-for-gdp/ (accessed 20 August 2021).

Gössling, S. (2020). Technology, ICT and tourism: From big data to the big picture. *Journal of Sustainable Tourism*, 29(5), 849–858.

Hasnat, I., and Steyn, E. (2021). The Curious Case of Bangladesh and Nepal: Tourism Advertising to Transform Country Image and Empower Developing Countries. In A. Hassan (Ed.), *Tourism Marketing in Bangladesh: An Introduction*. Abingdon: Routledge, pp. 299–313.

Haugen, J. (2020). How tech could help tourist destinations struggling in the wake of COVID-19. *FASTCOMPANY*. Retrieved from: www.fastcompany.com/90478303/how-tech-could-help-tourist-destinations-struggling-in-the-wake-of-covid-19 (accessed 11 August 2021).

Henderson, T. (2020). Coronavirus will slam states dependent on tourism. *PEW*. Retrieved from: www.pewtrusts.org/en/research-and-analysis/blogs/stateline/2020/03/16/coronavirus-will-slam-states-dependent-on-tourism (accessed 20 August 2021).

Hojeghan, S.B., and Esfangareh, A.N. (2011). Digital economy and tourism impacts, influences and challenges. *Procedia Social and Behavioral Sciences*, 19, 308–316.

Hospitality Technology (2021). *How are hotels adapting and innovating during COVID-19?* Retrieved from: https://hospitalitytech.com/how-are-hotels-adapting-and-innovating-during-covid-19 (accessed 20 August 2021).

India Brand Equity Foundation (2021). *Indian tourism and hospitality industry analysis.* Retrieved from: www.ibef.org/industry/indian-tourism-and-hospitality-industry-analysis-presentation (accessed 20 August 2021).

Islam, S.M.N, Al-Amin, M., and Akter, S.R. (2021). COVID-19 impact on regional tourism development in Bangladesh: A study of Cumilla District. *South Asian Journal of Social Studies and Economics*, 9(1), 50–58.

Jaipuria, S., Parida, R., and Ray, P. (2021). The impact of Covid-19 on tourism sector in India. *Tourism Recreation Research*, 46(2), 245–260.

Kim, D., and Kim, S. (2017). The role of mobile technology in tourism: Patents, articles, news, and motile tour app reviews. *Sustainability*, 9(11), 2082.

Lau, A. (2020). New technologies used in COVID-19 for business survival: Insights from the hotel sector in China. *Information Technology & Tourism*, 22, 497–504.

Madden, D. (2020). Ranked: The 10 countries most dependent on tourism. *Forbes*. Retrieved from: www.forbes.com/sites/duncanmadden/2020/04/02/ranked-the-10-countries-most-dependent-on-tourism/?sh=12f22e755939 (accessed 20 August 2021).

Magio, K.O. (2018). *Technological advancements and artificial intelligence can be a threat to tourism and hospitality.* Retrieved from: https://kennedymagio.com/2018/03/13/technological-advancements-and-artificial-intelligence-can-be-a-threat-to-tourism-and-hospitality/ (accessed 27 August 2021).

Mathur, S. (2021). Covid impact of tourism: 14.5 million jobs lost in 2020's Q1. *The Times of India*. Retrieved from: https://timesofindia.indiatimes.com/business/india-business/covid-impact-of-tourism-14-5-million-jobs-lost-in-2020s-q1-alone/articleshow/84573975.cms (accessed 20 August 2021).

Neuhofer, B., Buhalis, D., and Ladkin, A. (2013). A typology of technology-enhanced tourism experiences. *International Journal of Tourism Research*, 16(4), 340–350.

Organisation for Economic Cooperation and Development (2020). *Rebuilding tourism for the future: COVID-19 policy response and recovery.* Retrieved from: www.oecd.org/coronavirus/policy-responses/rebuilding-tourism-for-the-future-covid-19-policy-responses-and-recovery-bced9859/ (accessed 12 August 2021).

Pierce, J. (2021). COVID-19 has sped up hospitality technology. *Torrens University Australia*. Retrieved from: www.torrens.edu.au/blog/how-covid-19-has-sped-up-hospitality-technology#.YVYEui1h3Ay (accessed 20 August 2021).

PlugandPlay (2020). *What will happen to the travel industry after Coronavirus (and 6 startups that can help).* Retrieved from: www.plugandplaytechcenter.com/resources/impact-covid-19-travel-hospitality-industry-and-6-startups-can-help/ (accessed 20 August 2021).

Polas, M.R.H., Saha, R.K., and Tabash, M.I. (2021). How does tourist perception lead to tourist hesitation? Empirical evidence from Bangladesh. *Environment, Development and Sustainability*. DOI: https://doi.org/10.1007/s10668-021-01581-z.

Polland, J. (2014). A detailed look at how Americans travel within the US. *Business Insider*. Retrieved from: www.businessinsider.com/the-most-popular-us-states-for-tourism-2014-10 (accessed 20 August 2021).

Rogers, S. (2020). How virtual reality could help the travel and tourism industry in the aftermath of the Coronavirus outbreak. *Forbes*. Retrieved from: www.forbes.com/sites/solrogers/2020/03/18/virtual-reality-and-tourism-whats-already-happening-is-it-the-future/?sh=6c2da19828a6 (accessed 20 August 2021).

SelectUSA (2021). *Travel, tourism and hospitality spotlight. The travel, tourism and hospitality industry in the United States*. Retrieved from: www.selectusa.gov/travel-tourism-and-hospitality-industry-united-states (accessed 12 August 2021).

Sharma G.D., Thomas, A.,and Paul, J. (2021). Reviving tourism industry post-COVID-19: A resilience-based framework. *Elsevier Public Health Emergency Collection*, DOI: 10.1016/j.tmp.2020.100786.

Shetty, P. (2021). The impact of COVID-19 in the Indian tourism and hospitality industry: Brief report. *Journal of Tourism and Hospitality*, 10(1), 1–7.

Simpson, D. (2012). *Tourism vs. other industries: WTTC release comparisons for China and India*. Retrieved from: www.cabi.org/leisuretourism/news/22488 (accessed 14 August 2021).

Statista (2020). *Countries with the highest share of GDP generated by direct travel and tourism worldwide in 2019*. Retrieved from: www.statista.com/statistics/1100368/countries-highest-gdp-travel-tourism/ (accessed 20 August 2021).

Statista (2021). *Travel and tourism industry in the US. – statistics and facts*. Retrieved from: www.statista.com/topics/1987/travel-and-tourism-industry-in-the-us/. (accessed 14 August 2021).

US Travel Association (2020). *US Travel and Tourism Overview (2019)*. Retrieved from: www.ustravel.org/system/files/media_root/document/Research_Fact-Sheet_US-Travel-and-Tourism-Overview.pdf (accessed 20 August 2021).

US Travel Association (2021). *Covid-19 travel industry research. Travel recovery insights dashboard*. Retrieved from: www.ustravel.org/toolkit/covid-19-travel-industry-research (accessed 20 August 2021).

United Nations World Tourism Organization (2021). *Tourism and Covid-19 – Unprecedented economic impacts*. Retrieved from: www.unwto.org/tourism-and-covid-19-unprecedented-economic-impacts (accessed 26 August 2021).

Vidal, B. (2019). The new technology and travel revolution. *we are marketing*. Retrieved from: www.wearemarketing.com/blog/tourism-and-technology-how-tech-is-revolutionizing-travel.html (accessed 27 August 2021).

World Economic Forum (2021). *How virtual tourism can rebuild travel for a post-pandemic world*. Retrieved from: www.weforum.org/agenda/2021/05/covid-19-travel-tourism-virtual-reality/ (accessed 29 August 2021).

World Health Organization (2020). *WHO Director-General's opening remarks at the media briefing on COVID-19 – 11 March 2020*. Retrieved from: www.who.int/director-general/speeches/detail/who-director-general-s-opening-remarks-at-the-media-briefing-on-covid-19---11-march-2020 (accessed 20 August 2021).

World Travel and Tourism Council (2021). *Economic Impact Reports*. Retrieved from: https://wttc.org/Research/Economic-Impact (accessed 20 August 2021).

WorldAtlas (2018) *The most visited states in the US*. Retrieved from: www.worldatlas.com/articles/the-most-visited-states-in-the-us.html (accessed 10 August 2021).

Yehia, Y. (2019). The importance of tourism on economies and businesses. *globalEDGE*. Retrieved from: https://globaledge.msu.edu/blog/post/55748/the-importance-of-tourism-on-economies-a (accessed 18 August 2021).

Zambello, L. (2021). Hotel innovations and trends pushing the industry forward during the pandemic. *eHotelier*. Retrieved from: https://insights.ehotelier.com/insights/2021/02/24/hotel-innovations-and-trends-pushing-the-industry-forward-during-the-pandemic/ (accessed 20 August 2021).

Part VI
Cases from South America

16 Technology in Tourist Events

A Study of Rock in Rio Brazil from the Perspective of Its Stakeholders

Annaelise Fritz Machado, Bruno Sousa and Joice Lavandoski

Introduction

Cities see events as drivers of development for city marketing, as they provide opportunities for dynamization at the urban level (Lohmann and Dredge, 2012; Tseng et al., 2015). The events have a comprehensive role, working as a means of boosting the local economy, stimulating social inclusion, promoting culture and contributing to the generation of benefits for the city, since they are engines of development. With an audience of 700,000 people over seven days, Rock in Rio can be considered an activation platform for sponsoring and supporting brands, which increases the experiences they deliver to participants at every incarnation of the event. The new technologies recently introduced during Rock in Rio 2019 have changed the way organizers work, as well as the speed of task execution, the quality of the final product, companies' responsiveness and the overall operating costs. The way customer service is provided has also changed, as customers are more demanding now, they want quick returns and expect to live more effective and realistic experiences. The writing of this article is justified by the scarcity of scientific material related to the audience's technological experiences during the Rock in Rio Brazil 2019, which contributed to actively engaging them in the event and with the sponsoring brands/stakeholders. In this sense, the methodology used is descriptive research, presenting the Rock in Rio Brazil 2019 case as an object of study. The researcher Annaelise Machado went on a field trip to the City of Rock, where the concerts take place, in order to see what Rock in Rio sponsors and organizers have done with regard to technologies to engage the audience participating in the event. In order to compose this article, it was first necessary to build a solid theoretical foundation, and that was divided into three sections. In the first section, the approach concerned addressing the conceptual bases related to events and mega-events in the context of tourism; the second section pointed out the benefits of technologies and sustainability in events; and the third section addressed the technological resources included in Rock in Rio 2019 by the event's organizers and its stakeholders.

DOI: 10.4324/9781003271147-23

Events and Mega-Events in the Context of Tourism

Tourism is one of the most relevant and significant engines for the development and growth of cities' economies, in global terms (Sousa and Rodrigues, 2019). Year after year, there has been an increase in competition between tourist destinations (Sun et al., 2015; Matos and Barbosa, 2018). Therefore, it is not surprising that studies are carried out in the field of tourism, although they present different perspectives (Sousa and Simões, 2018). When analysing the economic dependence related to tourism, taking into account Gross Domestic Product (GDP), we found that Portugal has a dependence of 13.7%; Spain, 12.3%; France, 9.5%; Great Britain, 9%; Italy, 5%; Brazil, 3.7%; and the United States, 2.6% (Cruz, 2020).

With the COVID-19 pandemic, which started at the end of 2019 in China, and was declared a pandemic by the World Health Organization (WHO) in March of 2020, tourist activity has suffered a major negative impact with the closure of cities/countries to prevent the proliferation of the disease. "According to calculations by the United Nations World Tourism Organization (UNWTO), international tourist flows had a 22% drop in 2020, as well as a 20% to 30% drop in revenue generated in the sector" (Cruz, 2020). According to Sayuri (2021), the estimated loss was US$1.1 trillion by the end of 2020. It is estimated that the sector may start recovering in 2023.

As a segment of tourism, events are also being heavily impacted, and the return to regular activities is very slow. Divided into numerous categories, events are distinguished by their complexity, size and scope, or according to their concept or purpose. Their main attributes are usually the gathering of people for common purposes, spatiotemporal selectivity in their realization, and an unscheduled routine. They can be represented in different natures, dimensions, reach and duration (Molina, 2016).

In the twenty-first century, companies from different sectors have included events as part of their activities in order to reduce advertising expenses, generate sales of products/services that promote profits, and to work on their branding/image, marketing and institutional return. Mega-events, which are the focus of this research, are true spectacles and are configured as urban strategies of the government allied with private initiative, and entail a profusion of urban megaprojects (involving various construction activities) that directly impact the urban structure of cities (Sposito, 1991; Molina, 2016).

The various elements linked to mega-events are directly involved in the proposed urban development and tourism segmentation, as well as the creation of infrastructure and facilities for their realization, which can and should be extended to the local community in the search for beneficial results for all actors involved. Thus, a mega event is considered successful when it promotes the dissemination of a positive image of the host city internationally and allows for a positive legacy for the local community and tourists, minimizing negative impacts (Silva et al., 2016).

By March 2020, there were great evolutions in the events market, and technology became a fundamental tool, offering companies greater reach with regard to the market. Most of this change was due to the new type of human that the "Age of Access" has brought about. Today, young people (especially) feel much more comfortable in electronic environments than in traditional ones (Rifkin, 2001; Krum, 2010; Montenegro, 2017). Thus, the themes of technology and technological innovation in events are seen in the context of entertainment management, as they provide tools that help to facilitate the work of event organizers, a point which will be discussed below.

The Benefits of Technology and Sustainability in Events

When discussing technology in events, it is necessary to present a description of the word "innovation", a term that means "connections, interactions and influences of varying degrees – including relationships between companies, research centers and the government" (Caldas, 2001; Tidd et al., 2008; Lorenzetti et al, 2012). The Brazilian Innovation Law defines innovation as "introduction of novelty or improvement in the productive or social environment that results in new products, processes or services" (Presidência da República Secretaria-Geral Subchefia para Assuntos Jurídicos, 2004; Lorenzetti et al., 2012).

Currently, when discussing events, there is a concern to make the organizers' lives easier, allowing for the insertion of tools that help them with organization and planning, which, consequently, provides a better experience and greater engagement of the audience (Costa et al., 2013; Matos et al., 2016). Therefore, technology has become an important resource used in this process. By using it in events, it is possible to create a strategy that is more in line with what the audience wants.

Technology provides great contributions when it comes to events: it offers greater interaction with the audience, avoiding or reducing queues and delays, helping to minimize errors that may cause negative effects, offering registration through platforms, allowing people to confirm their presence through smartphones, generating QR codes that can be scanned upon arrival at the event, managing participants and accommodations, sending certificates, transmitting the event via streaming and providing greater reach of dissemination (Sinarta and Buhalis, 2018). The positive results include reaching more people, reducing costs and greater interaction with participants (Costa et al., 2013).

Other technological innovations used in the field of events include biometrics, such as assessment of things such as the participants' heartbeat or the recognition of facial expressions; robotics; and apps, which facilitate interaction between participants and their engagement. Also, there are the technologists, who are in charge of searching for solutions and providing creative means of using technology to improve the experience of the participants (Grupohel Eventos Digitais, n.d.).

Considering that societies are driven by the latest technology, it promotes the movement of people and the search for products/services, in addition to creating new challenges and generating greater demands (Boas, Sousa and Fernandes, 2019; Boas and Sousa, 2019; Neuhofer, Buhalis and Ladkin, 2014). Among these options, there is the desire to live innovative and sustainable experiences.

Specialists in the field of tourism know that events generate not only positive, but also negative impacts. Therefore, in light of the development of environmental awareness on the part of consumers, as well as event organizers, with regard to making the planet cleaner, environmental conservation actions are carried out for events (Macdowell and Silberberg, 2010). The sustainability of an event promoting company permeates a set of corporate actions that simultaneously propose social integration, economic equilibrium and environmental responsibility (Mazza et al., 2014). It must demonstrate "in its planning and implementation process, the environmental impacts, from the measurement to the disposal of the waste generated in its execution, the consumption of electricity and the neutralization of carbon dioxide (CO_2) emissions through planting of native trees" (Matias, 2011, p. 200).

Sustainable events improve the image of the organizing company, increase the number of followers (because they enjoy environmental responsibility), decrease costs, increase the organizers' profits, and make people happier, because they feel committed to minimizing harm to the planet. The next section focuses on the technological resources used in Rock in Rio 2019, which influence the engagement of the audience.

Technological Resources and Sustainable Actions included by the Organizers of the Event and Its Stakeholders in Rock in Rio 2019

Rock in Rio was created in 1985 by businessman Roberto Medina right after the end of Brazil's military dictatorship. Since its first incarnation, and until today, it has been held in the city of Rio de Janeiro, Brazil and in some others (Rock in Rio, n.d.a; Costa and Igreja, 2019). Since its inception, the event has been recognized as the largest music festival in the world. It toured the world for ten years and then returned to Brazil in 2001. Since its first incarnation in Lisbon, Portugal, the event has been held either in Brazil or in some other major city, such as Madrid and Las Vegas (Silva and Lima, 2014). Rock in Rio's values are to make a difference in the world, therefore, it supports various social and environmental causes (Época, 2017). Its motto is "For a Better World", which can only be accomplished with everyone's participation.

The gradual transformation of technological means has opened new perspectives and different ways of communicating and presenting the tourist product/service. Those mechanisms have shattered the structure of the tourist chain, making it possible to reduce the distance between the organizer and the consumer of the tourist product/service (Macedo and Sousa, 2019). It is noteworthy how technologies have made a strong contribution to the events sector and allowed consumers to have a faster, broader and more comprehensive knowledge of tourist offers in different parts of the globe (Buhalis et al., 2012).

In addition, investing in technology produces good results for event producers, as well as for the stakeholders involved, since it produces a significant increase in sales, improvements in the relationship between the consumer and the brand, as well as a direct impact on brand image and greater profitability. As a result of that scenario, thinking about the best experience for their regular/consumer audience, the entire Rock in Rio 2019 festival was organized with the help of various technological resources brought about by the event organizers and the stakeholders/sponsors.

Technological resources used in Rock in Rio by the event organizers

Rock in Rio generated 100 TB of the Internet data traffic during the seven days of the event, 35% more than the entire 2014 World Cup. To understand what this volume means, this number corresponds to 2 million hours of streaming music or 28 million high-resolution photos (Rosa, 2019).

The ticket to enter the event was a uniquely designed bracelet with a chip that replaced the paper ticket and the card used in previous incarnations. The bracelet could only be worn on the day of the event, as it had a tamper-proof seal with technology for electronic validation in entrance control. With the bracelet in hand, it was important to visit the official Rock in Rio website to register it. That same bracelet could be used at the food and beverage stalls and anything purchased was registered on it. Purchases were deducted from the bracelet credit (Crossetti, 2019). The bracelets were also used to control the lines for the rides.

Anyone who wanted to download the Rock in Rio app on their cell phone could locate themselves within the City of Rock/Olympic Park (Crossetti, 2019). The event had cameras to monitor, in real time, the queue of the arenas where games and attractions took place, such as the Fuerza Bruta group (Rosa, 2019). Technology was also the festival's ally to monitor the audience. For the first time, Rock in Rio had several facial recognition cameras spread across the Olympic Park, capturing more than 700,000 people who passed through the Olympic Park over the seven days of the event. Moreover, security guards around the event used cameras attached to their uniforms, with real-time transmission, to increase the security of the audience enjoying the festival. In addition, drones, thermal cameras and a system that can identify items that may pose a security risk were used (Grinberg, 2019).

Visually impaired people who visited the attractions had the unprecedented experience of circulating and interacting with the City of Rock in a high level of accessibility. By downloading the Veever app for free, blind people or people with different levels of visual impairment had access to spatial information that was transmitted by voice command (Luz, 2019).

In the restrooms, the turnstiles had sensors to measure the flow of people. This information was available in the event app. Thus, people could see which bathroom was less full (Rosa, 2019). There was also news in the lost and found segment. If a person lost a document or a cell phone, they could check the app, since ID cards would be scanned and objects would have their characteristics described there (Rosa, 2019).

To further engage the people who attended the event, a company called Medina implemented a chatbot, "Roque", to interact with Facebook users (Crossetti, 2019). While the concerts took place on the stage, social networks were being filled with posts about the event. Twitter, in particular, was responsible for more than 4 million mentions as early as on the first weekend of the event, with several hashtags (Crossetti, 2019).

Technological resources used by the stakeholders/sponsors at Rock in Rio

As for the technological innovations used at the event, the highlight was the industrial robot Yumi, a kind of bartender inside a futuristic bar. That area was created in partnership with Heineken and ABB. The robot served water to remind consumers of the importance of responsible drinking (Rosa, 2019). In addition, the hashtag #HeinekenPlay was created. All you had to do was to tweet this hashtag with the name of the song you would like to hear for the request to appear on the screen of the World Stage (Palco Mundo), so that the band could see it. The audience that used the hashtags #AquaRiR and #ConhecerParaConservar got tickets to the Rio de Janeiro Marine Aquarium (AquaRio) and the Christ the Redeemer statue at the AquaRio booth at Rock in Rio (Crossetti, 2019). Since its implementation, the Rock in Rio Operation has had a robust infrastructure, supported by more than 360,000 km of optical fibre, 500 access points to the Oi Wi-Fi network in Brazil, and more than 56 kilometres of fibre installed exclusively for the event. Oi recorded 164.94 TB of traffic during the event, with more than 10.8 million Wi-Fi connections, with a peak of 18,474 simultaneous connections using Oi Wi-Fi and 5G technology. The partial data consumption over the seven days of the festival reached 82.85 TB (Rock in Rio, n.d.b).

Mixed reality activities were also offered, in addition to a "virtual bike ride" through a videowall showcasing the Aterro do Flamengo (Flamengo Park). The bike was connected to Oi 5G and it responded to the user's stimulus – the faster they pedalled, the faster the playback was in the video in front of them (Rosa, 2019).

Biometrics was used to access the 24 Horas ATMs in Rock Street Asia. Although the project belongs to Itaú, customers of various financial institutions were able to withdraw money only using their bank data, PIN and fingerprint validation, without the need for a physical card (Rosa, 2019).

For the first time, Rock in Rio used the velodrome area to transform it into an area called NAVE (*Nosso Futuro é Agora* [Our Future is Now]) – a co-creation with Natura. The experience materialized the discourse that in order to build the future it is necessary to change the present, using projections, scenographic imagery and special effects, music and smells to stimulate a view of the future (Franchini, 2019). The games area had 163 game stations – including the first 5G mobile in Brazil – three game releases (Contra Rogue Corps, Call of

Duty: Mobile, and Ghost Recon Breakpoint) and a space for 12 independent developers from Rio de Janeiro (Rock in Rio, n.d.b).

Linking the event to social networks, Skol transformed a person's Messenger QR Code into a pin to make the exchange of messages easier (Crossetti, 2019). In addition to brands, media companies also invested heavily in Rock in Rio. This was the case of Rede Globo, which set up Globo and Globoplay stalls in the City of Rock on the first weekend of the festival (Propmark, 2019).

There were more innovations in front of the Palco Mundo (World Stage). In Globoplay's space, the audience could watch the main concerts of the event as though they were close to the singers or bands. That was made possible with the use of virtual reality (VR) stations and a camera system that filmed in 360 degrees, and 8K (ultra-high definition resolution) and HDR (a feature that raises the quality with more brightness and contrast) (Rosa, 2019). Among its stakeholders, Rock in Rio 2019 had Filtr, Sony Music Brasil's entertainment platform. Through an online game, Filtr Game, it was possible to accumulate Filtr Coins, the virtual coins of the game, which could be exchanged for products and/or benefits in the Filtr Store (Diário do Rio, n.d.).

In partnership with Otima, Trident brought in the first information panel with artificial intelligence based on image recognition. The technology provided a map of the event, resources to charge cell phone batteries and Wi-Fi connection (Levin, 2017). To make life easier for those who wanted to find a soul mate in City of Rock, Tinder prepared a special feature. That was Festival Mode, a feature within the app designed to facilitate interaction between people who went to enjoy Rock in Rio. The festival enabled a match between people who attended the event, even before it started (Diário do Rio, n.d.).

Tattoos have also entered the digital age. It was possible to get a tattoo with a theme related to the world of music from a skin scan made with a device that resembles a cell phone, in Submarino's space. Organizers guaranteed that the tattoo would vanish within three days (Rosa, 2019).

As for the sustainable actions developed by Rock in Rio, there was the Sustainability Plan, with social, environmental and economic measures, and Waste Management (reduce, recycle and reuse 145 tons). In its last incarnation, they recycled and reused garbage in the making of deodorant caps for Natura. The sponsoring stakeholders gave prizes for the best sustainable practices with the Rock in Rio Atitude Sustentável (Sustainable Attitude), with part of their income donated to social projects (Rock in Rio, n.d.c).

Other similar actions were the calculation of the carbon footprint; Amazonia Life – concerned with sustainable attitudes – the planting of 1 million trees in deforested areas at the source of the Xingu River; 4 million trees planted in the Amazon; lower consumption of water and electric power at the event (solar trees – to recharge cell phones with clean energy) (Rock in Rio, n.d.c). In 2013, the event received the ISO 20121:2012 international certification, due to the sustainable actions developed at the event (Schuchmann and Schuchmann, 2019). The challenge was to keep everyone involved committed to sustainable actions and to not act alone (Rock in Rio, n.d.d).

Final Considerations

Producing events has always been considered one of the most stressful activities with regard to planning and organization, compared to other professions in the events market. The addition of new technological tools has made the process more agile and efficient. The twenty-first century is marked by the frequent use of technology in events, in order to facilitate the planning, organization and execution process, avoiding rework and unnecessary waste of time by those involved in the production. Technological innovations have brought about many contributions to the field of events, such as: better digital management, mobile engagement, virtual or augmented reality, integrated platforms, innovative experiences, gamification and tools for social networks, among other innovations.

During the Rock in Rio 2019, the engagement and dialogue between brands and their products with the consumer audience changed. It was a moment of delivery of experiences, but also of testing technological innovations. The Internet and technological innovations during Rock in Rio have become protagonists of this new scenario, contributing to the closer connection of brands with their customers, fostering better engagement in the event. In addition, sustainability during the event became a reality when they included the Management Plan based on the pillars of sustainable development (Social, Economic, Environmental), and highlighted green marketing.

With the needs of the Rock in Rio audience met, new fronts must be thought of and new experiences offered. From an interdisciplinary perspective, this article presents inputs into the fields of tourism (e.g., events and culture), marketing (e.g., consumer engagement) and technological development of tourism and sustainability. Future research should be carried out in the form of quantitative studies, through the administration of questionnaire-based surveys, in order to understand the impact of the influence of technology on the level of symbolic and affective relationship with a particular place or tourist event (e.g., structural equation models).

References

Boas, V.V., and Sousa, B. (2019). The role of service quality in predisposition for Portuguese online commerce. *Quality – Access to Success*, 20, 12–17.

Boas, V.V., Sousa, B., and Fernandes, P.O. (2019). *Trust and perceived risk in consumer behaviour: A preliminary study in electronic commerce*. Retrieved from: https://biblio tecadigital.ipb.pt/bitstream/10198/21509/1/Trust%20and%20Perceived%20Risk. pdf (accessed 7 August 2021).

Buhalis, D., Darcy, S., and Ambrose, I. (2012). *Best Practice in Accessible Tourism: Inclusion, Disability, Ageing Population and Tourism*. Bristol: Channel View Publications.

Caldas, R.A. (2001). A construção de um modelo de arcabouço legal para ciência, tecnologia e inovação. *Parcerias Estratégicas*, 11(6), 5–27.

Costa, A., and Igreja, A. (2019). *Rock in Rio: a arte de sonhar e fazer acontecer*. São Paulo: Editora Gente.

Costa, C., Panyik, E., and Buhalis, D. (Eds.). (2013). *Trends in European Tourism Planning and Organisation*. Bristol: Channel View Publications.

Crossetti, M.C. (2019). Rock in Rio: mapa, dicas e apps para usar na Cidade do Rock. *tecnoblog*. Retrieved from: https://tecnoblog.net/308985/rock-in-rio-dicas-e-apps-para-usar-na-cidade-do-rock/ (accessed 10 August 2021).

Cruz, R. de C.A. Da. (2020). Impactos da pandemia no setor de turismo. *Jornal da USP*. Retrieved from: https://jornal.usp.br/artigos/impactos-da-pandemia-no-setor-de-turismo/ (accessed 8 August 2021).

Diário do Rio (n.d.). *Festival Mode On, o espaço do Tinder para 'matches reais' no Rock in Rio*. Retrieved from: https://diariodorio.com/festival-mode-on-o-espaco-do-tinder-para-matches-reais-no-rock-in-rio/ (accessed 10 August 2021).

Época (2017). *Rock in Rio apoia ações sociais e ambientais*. Retrieved from: https://epoca.globo.com/Especial-Publicitario/Rock-in-Rio/noticia/2017/09/rock-rio-apoia-acoes-sociais-e-ambientais.html (accessed 6 January 2021).

Franchini, G. (2019). Rock in Rio 2019: Atração "Nave – Nosso Futuro é Agora" promove sessão especial. *Universo do Rock*. Retrieved from: https://universodorock.com/rock-in-rio-2019-atracao-nave-nosso-futuro-e-agora-promove-sessao-especial/ (accessed 1 August 2021).

Grinberg, F. (2019). Rock in Rio estreia sistema de câmeras de reconhecimento facial. *O Globo Rio*. Retrieved from: https://oglobo.globo.com/rio/rock-in-rio-estreia-sistema-de-cameras-de-reconhecimento-facial-23980877 (accessed 10 August 2021).

Grupohel Eventos Digitais (n.d.). *A Revolução da Tecnologia nos Eventos Corporativos*. Retrieved from: www.grupohel.com/produtos-e-servicos/tecnologia-em-eventos/ (accessed 10 August 2021).

Krum, C. (2010). *Mobile Marketing: Finding Your Customers No Matter Where They Are*. London: Pearson Education, Inc.

Levin, T. (2017). As ativações das marcas no Rock in Rio. *Meio&mensagem*. Retrieved from: www.meioemensagem.com.br/home/marketing/2017/09/12/as-ativacoes-das-marcas-no-rock-in-rio.html (accessed 11 August 2021).

Lohmann, G., and Dredge, D. (2012). *Tourism in Brazil: Environment, Management and Segments*. Abingdon: Routledge.

Lorenzetti, J., Trindade, L. de L., Pires, D.E.P. De., and Ramos, F.R.S. (2012). Tecnologia, inovação tecnológica e saúde: uma reflexão. *Texto Contexto Enferm, Florianópolis, Abr-Jun*, 21(2), 432–439.

Luz, S. (2019). Tecnologia e Inclusão para Pessoas com Deficiência Visual no Rock in Rio 2019. *Voicers*. Retrieved from: www.voicers.com.br/tecnologia-e-inclusao-para-pessoas-com-deficiencia-visual-no-rock-in-rio-2019/ (accessed 10 August 2021).

Macdowell, D., and Silberberg, C.P. (2010). *Gestão ambiental e responsabilidade social em eventos*. In A. Jr. Philippi and D.V. de M. Ruschmann (Eds.). *Gestão Ambiental e Sustentabilidade no Turismo*. Barueri: Manole, pp. 735–754.

Macedo, C., and Sousa, B. (2019). A acessibilidade no etourism: um estudo na ótica das pessoas portadoras de necessidades especiais. *Pasos. Revista de Turismo y Patrimonio Cultural*, 17(4), 709–723.

Matias, M. (2011). *Planejamento, organização e sustentabilidade em eventos culturais, sociais e esportivos*. Barueri: Manole, p. 200.

Matos, M.B.A., and Barbosa, M.D.L.A. (2018). Autenticidade em Experiências de Turismo: proposição de um novo olhar baseado na Teoria da Complexidade de Edgar Morin. *Revista Brasileira de Pesquisa em Turismo*, 12(3), 154–171.

Matos, M.B.A., Matos, B.G., and Barbosa, M.D.L.A. (2016). Agente versus estrutura no modelo multidimensional-reflexivo: estudo de caso de uma organização hoteleira do tipo cama café. *Revista de Turismo Contemporâneo*, 4(1), 138–155.

Mazza, C., Isidro-Filho, A., and Hoffmann, V.E. (2014). Capacidades dinâmicas e inovação em serviços envolvidas na implementação e manutenção de práticas de sustentabilidade empresarial. *Revista de Administração e Inovação*, 11(1), 345–371.

Molina, F.S. (2016). A produção da "Paris dos Trópicos" e os megaeventos no Rio de Janeiro no início do século XX. *Finisterra, LI*, 102, 25–45.

Montenegro, J.L.Z. (2017). Revisão Bibliográfica sobre Fatores de Impacto no Mobile Commerce. *Revista de Empreendedorismo, Inovação e Tecnologia*, 4(1), 77–91.

Neuhofer, B., Buhalis, D., and Ladkin, A. (2014). A typology of technology-enhanced tourism experiences. *International Journal of Tourism Research*, 16(4), 340–350.

Propmark (2019). *Rock in Rio: tecnologia conecta conteúdo e público nos estandes da Globo*. Retrieved from: https://propmark.com.br/rockinrio-2019/rock-in-rio-tecnologia-conecta-conteudo-e-publico-nos-estandes-da-globo/ (accessed 5 August 2021).

Rifkin, J. (2001). *A era do acesso*. São Paulo: Makron Books.

Rock in Rio (n.d.a). *História do Rock in Rio*. Retrieved from: https://rockinrio.com/rio/pt-BR/historia/ (accessed 17 January 2021).

Rock in Rio (n.d.b). *Rock in Rio fecha edição de 2019 com números impressionantes*. Retrieved from: https://cdn.rockinrio.com/wp-content/uploads/2020/03/apv_rock-in-rio-fecha-edicao-de-2019-com-numeros-impressionantes_16-10.pdf (accessed 10 August 2021).

Rock in Rio (n.d.c). *Rock in Rio anuncia ganhadores do Prêmio Rock in Rio Atitude Sustentável 2019*. Retrieved from: https://cdn.rockinrio.com/wp-content/uploads/2020/03/rock-in-rio-2019_-rock-in-rio-anuncia-ganhadores-do-premio-rock-in-rio-atitude-sustentavel-2019_.pdf (accessed 3 February 2021).

Rock in Rio (n.d.d). *Plano de Sustentabilidade Versão 1.1*. Retrieved from: http://cdn.rockinrio.com.br.s3.amazonaws.com/manual_uploads/Rock_in_Rio_2019_Plano_de_Sustentabilidade_versao_1_1.pdf (accessed 1 January 2021).

Rosa, B. (2019). Rock in Rio 2019: Edição mais hi-tech da História tem robô-garçom. *O GLOBO*. Retrieved from: https://oglobo.globo.com/cultura/rock-in-rio-2019-edicao-mais-hi-tech-da-historia-tem-robo-garcom-23977449 (accessed 7 August 2021).

Sayuri, J. (2021). O fenômeno do travel shaming. E o turismo na pandemia. *NEXO*. Retrieved from: www.nexojornal.com.br/expresso/2021/04/16/O-fen%C3%B4meno-do-travel-shaming.-E-o-turismo-na-pandemia (accessed 7 August 2021).

Schuchmann, C., and Schuchmann, B.M. (2019). Um estudo sobre sustentabilidade em eventos: Rock In Rio. *Revista Científica Multidisciplinar Núcleo do Conhecimento*, 03, 69–77.

Silva, A.C. da., Braga, D.C., and Romano, F.S. (2016). Megaeventos e turismo: umm estudo bibliométrico dos periódicos brasileiros de Turismo. *Turismo - Visão e Ação*, 18(3), 633–659.

Silva, H. da L., and Lima, M. do N. (2014). De maior festival de Rock no Brasil para o maior Festival de Música do mundo: uma análise da repercussão do Rock in Rio entre fãs, apreciadores e antifãs no Twitter. *VIII Simpósio Nacional da ABCiber*. São Paulo: Unifoa.

Sinarta, Y., and Buhalis, D. (2018). Technology Empowered Real-Time Service. In B. Stangl and J. Pesonen (Eds.), *Information and Communication Technologies in Tourism 2018*. Cham: Springer, pp. 283–295.

Sousa, B., and Rodrigues, S. (2019). The role of personal brand on consumer behaviour in tourism contexts: The case of madeira. Enlightening Tourism. *A Pathmarketing Journal*, 9(1), 38–62.

Sousa, B., and Simões, C. (2018). An Approach on Place Attachment, Involvement and Behavioural Intentions in Iberian Marketing Contexts: The Case of Galicia-North Portugal Euroregion: An Abstract. P. Rossi and N. Krey (Eds.), *Finding New Ways to Engage and Satisfy Global Customers, Developments in Marketing Science: Proceedings of the Academy of Marketing Science*, https://doi.org/10.1007/978-3-030-02568-7_165

Sposito, M.E.B. (1991). O centro e as formas de expressão da centralidade urbana. *Revista de Geografia*, São Paulo, UNESP, 10, 1–18.

Subchefia para Assuntos Jurídicos (2004). *Lei nº 10973. Dispõe sobre incentivos à inovação e à pesquisa científica e tecnológica no ambiente produtivo e dá outras providências.* Brasília (DF): Congresso nacional.

Sun, M., Ryan, C., and Pan, S. (2015). Using Chinese travel blogs to examine perceived destination image: The case of New Zealand. *Journal of Travel Research*, 54(4), 543–555.

Tidd, J., Bessant, J., and Pavitt, K. (2008). *Gestão da Inovação.* São Paulo: Bookmann.

Tseng, C., Wu, B., Morrison, A.M., Zhang, J., and Chen, Y.C. (2015). Travel blogs on China as a destination image formation agent: A qualitative analysis using Leximancer. *Tourism Management*, 46, 347–358.

17 The Sacred in Cyberspace

The Taper of Our Lady of Nazareth Religious Event and Technology Application in the (Re)Construction of Territorial and Touristic Identities in Belém do Pará, Brazil

Annaelise Fritz Machado, André Luiz Lopes de Faria and Gabriela Oliveira Rodrigues

Introduction

Tourism, considered one of the sectors that significantly and globally drives the culture and economy of a given country, is one of the engines that contribute to local development and establish important social bonds that energize localities and control seasonality. Event Tourism, one of its segments, has become a great ally to locations that wish to develop socially, culturally and economically, by attracting people who leave their home towns and head to a destination city to participate in events of many different sorts. When dealing with events, it has been seen that religious ones attract people who, motivated by faith, participate in sacred rituals, mysteries of faith, pilgrimages, evangelization-related activities, or the worship of a saint, making use of the tourism infrastructure of the place visited as support. In this context, religious tourism emerges as a source of employment, driving various sectors of the economy, such as hotels and catering, in addition to stimulating the purchase of religious and handicraft items. This form of tourism is found in several destinations and is based on different religious beliefs.

The Círio de Nazaré (Taper of Our Lady of Nazareth) event in the city of Belém do Pará is considered one of the largest religious events in Brazil, drawing more than 2 million people. It needs to (re)build, migrating from its physical spaces to the social networks, with the incorporation of religious rituals in cyberspace. When dealing with cyberspace, one thinks of a digital space provided by the Internet and made up of interconnected computers, and understood as a space for the exchange of information in contemporary culture. This chapter asks what types of cyberspace technologies contribute to the re(construction) of territorial and touristic identities in Belém do Pará?

The first section deals with religious tourism and religious events; the second section deals with real territories and virtual territorialities emphasizing technologies in the (re)construction of territorial and tourist identities, and finally,

DOI: 10.4324/9781003271147-24

the third section on the sacred in cyberspace having as object of study the religious event Círio de Nazaré, presenting the migration from the event to the virtual one, and the emergence of the 360 degrees Círio – Virtual Círio Museum, Official Page, Facebook and WhatsApp and KD a Berlinda.

Religious Tourism and Religious Events

Nowadays, tourism is characterized as a global phenomenon, responsible for leveraging economies and organizing social, cultural, environmental and spatial practices that reconfigure local realities (Fragelli et al., 2019). Niche tourism or segment tourism is defined by the reasons for the trip and by the aspects of the destinations, in addition to the psychological, cultural or professional factors inherent to the individual, which lead to a growing heterogeneity in the level of tourist demand, fostering new types of tourism in different circumstances (Sousa and Simões, 2010).

Brazil is a tourist destination that deals with religious trips, where pilgrims, motivated by faith, by the sacred, by devotion to a saint, leave their home cities and go to destination cities, looking to visit churches/temples and participating in religious festivals (eventization), driving what we call Religious Tourism.

Religious Tourism is that which is carried out in places involving religiosity, whether the tourist is a believer, or not, of the religion in question. In different parts of the world, people motivated by the sacred promote a new reordering of the traditional geography of the faith, that is, a new spatiality and territorial expansion of religion (Nascimento and Souza, 2019).

Bauman (2008) defends the new social arrangements of "post-modernity" and highlights liquid modernity, where values are set aside and give way to the reproduction of a consumer-oriented way of life. "There is a relationship between religion (form) and consumption/trip/travel (content), in which the pilgrim tradition is measured by tourism" (Silveira, 2007: 41). Religion, as a component of culture, is embedded in and influences people's lives, with regard to clothing, eating and drinking, relating and politicizing (Kocyìgìt, 2016; Nyaupane et al., 2015).

Silveira (2007) states that religious tourism consists of visiting places considered sacred, using the tourist support infrastructure. It is noteworthy that this travel brings positive economic returns to the places where they are found, since consumer practices are common in these places.

The consumption provided by the tourist activity must be carried out from studies and analyses that provide information/data about the territory, which will guide the planning and territorial management processes, maximizing the positive effects of this use and minimizing the negative ones.

Griffing and Raj (2017) investigate the reason for the expansion of faith-motivated travel in recent years, as a trend towards secularization and a decline in the frequency of religious adherence is perceived worldwide.

In the historical context of tourism, there is a clear relationship with religion, when it comes to human migrations, for a variety of reasons, especially those of a religious nature and land exploration (Kocyìgìt, 2016). According to data from the Brazilian Ministry of Tourism (Mtur), an influx of around 14 million people, motivated by faith, is estimated in several Brazilian regions, which represents travel within the territory, of about 7% of the population, attracted by religious/ spiritual aspects, devotion to a saint, thanksgiving for grace received (Ministério da Cultura e Turismo, 2015; Aragão, 2014).

Nascimento and Souza (2019) point out, in 2017, the Catholic sanctuary of Aparecida do Norte, the festivities of Círio de Nazaré (Taper of Our Lady of Nazareth), the pilgrimages of Juazeiro do Norte and Canindé; and the enactment of the Paixão de Cristo de Nova Jerusalém (Passion of Christ of Nova Jerusalem) in Pernambuco, together and added to other destinations, generate a business of US$4.4 billion resulting from 20 million national trips per year (Nascimento and Souza, 2019).

Thus, the way to attract people and drive places economically is to hold events. As diverse as they are, they contribute to tourist movement, develop the tourist trade, control seasonality, encourage cultural exchanges and bring people together. Religious festivities are set up as events, independent of the religion to which they belong, and they promote beliefs, values, doctrines, devotions and popular manifestations in favour of a religion. In this context, they "draw" people and provide moments of modification of the space in which they are found, giving them meaning, qualifying them through their characteristics and transforming them into a unique place where the population's beliefs are the object of differentiation (Saraiva, 2010).

The growth of events in the religious area is a trend towards eventization of faith, reflected in the development of new models of engagement between church congregations and visitors and by church management and operational teams (Dowson, 2018, 2021).

The transition from a church to an event venue has raised questions, as this action has been promoted by some religious institutions, with the aim of acquiring financial resources for various purposes. However, the risk that this action can pose to religious spaces, regarding their use, regarding the loss of cultural and religious heritage, and the accelerated wear and tear of its assets has been questioned (Dowson, 2018).

During the COVID-19 pandemic, which began in March 2020 and continues with no end in mid-2021, religious environments suffered a strong impact with the prohibition of holding religious events in their spaces, as public health protocols made the risk of crowding people in closed, or even open spaces, clear. Given this reality, the use of digital media for religious practices became undeniable. There was a rupture from traditional events developed in person, to digital ones, and an accelerated use of streaming, lives, virtual visits, which will be presented throughout this chapter. Is this the new reality?

Real Territories and Virtual Territories

Tourism is a socio-spatial-cultural phenomenon of great symbolic value to the subjects who practise it and to the subjects who live in the place where it is practised (Castrogiovanni, 2004). It is in this so-called tourist space, that the elements that attract tourists are found, understood as attractions, as well as all the basic infrastructure, support and tourism that contribute to the smooth operation of the activity.

Based on Silveira's (2007) discourse, this space, constituted by the dialectical movement between form and content, "in the touristification processes, result in tourism territories, which are defined as real territories or tourist destinations" (Fratucci, 2014: 92). Visitors, in their stops at tourist destinations, appropriate spaces, settling on some, ignoring others and terrifying themselves in a reticular logic. "The other social agents involved by him and for him (tourists) are also creating their territories, which overlap and compose, denoted tourism territory" (Fratucci, 2014: 92).

People create bonds with places, due to their various movements in space, which become self-defining, relational and historical, to the extent that they contextualize people's relationships and enable the construction of their social life (Augé, 2017; Vilhena, 2002; Vilhena and Novaes, 2018).

The territorial format of tourist places is not, in and of itself, responsible for attracting visitors, they depend on the conceptions of territorial resources made available to the visitors, as well as on their image. Territory appears as a technique in a materialist vision, something to be taken and enjoyed by a group as a means of altering space (Góis, 2020).

According to Saquet (2009: 87), "territoriality takes place at different spatial scales and varies over time through power relations, circulation and communication networks, domination, identities". Meanwhile, thinking of the city of Belém do Pará, having the religious event Círio de Nazaré as a product of the city, territoriality goes beyond a political dimension, and is associated with social, economic, cultural and environmental relations, and the relationship of these dimensions with people and their spaces, how they organize themselves and give meaning to the place where they are inserted.

Haesbaert (2004) and Silva and Tourinho (2017) highlight these relations in their theoretical foundations, and add power relations, indicating where the effective territorial control of a particular group, people or nation goes, in addition to pointing out the existence of other territorialities.

When dealing with virtual territoriality in post-modernity, non-place becomes the perfect word for this time, intense individualism, loss of identity, excess of activities in a short space of time, the amount of events and the superabundance of space (Sá, 2014; Augé, 2017).

Lévy (1996), a student of the information age, says that in the past, the real (physical) territory had a certain connection with geographic location. With virtual territoriality, the location takes on a new look, new configurations, a possibility of being in several places at the same time, or nowhere. A place that spreads around the world, where you do not require the return of your gaze in terms of a face-to-face meeting, but there is a protective screen that filters, a place in the universe called clouds, goods and information: where you read emails, where one buys, plays, trains, speaks and concentrates using the most different resources (Augé, 2017; Vilhena and Novaes, 2018).

Virtual territoriality is also composed of space, but not a delimited geographic space, real, physical with borders, but a virtual space, which spreads around the world, becoming multidimensional, with a materiality of machines connected to each other, forming network territories, where social relationships are established through remote communities.

In this sense, as social relations begin to be submitted to virtual environments, formed by a contingent of information in a locus dissociated from geographical and physical space, it is shown that the notion of territory and territoriality, pre-viously only thought about with regard to the material, can also be submitted to these virtual environments, since it is related to a trustworthy space for the construction of culture and the establishment of human relations, even if it is not directly related to geographic space, as they are intangible, immaterial and meta-physical information spaces (Pimentel, 2010).

It is noted in the author's speech that, in different parts of the world, the interconnection of people will exist, as well as social relations. When dealing with the (re)construction of identities, the deterritorialization of physical places to territorialize virtual places is cited. In this dynamic, it is worth thinking about the emptying of physical spaces, the impoverishment of face-to-face cul-tural exchanges, the reduction of gains for the tourist trade, on the other hand, there is an increase in the use of digital media, greater virtual interactivity. This (re)construction was initially necessary to make the event more digital, later due to the COVID-19 pandemic, which made it impossible for the event to take place in person for the first time, provided the faithful with resources to keep the event alive, even if virtually, encouraging continued devotion to Our Lady of Nazareth.

Technologies in the (re)construction of territorial and tourist identities

Talking about (re)construction refers to the idea of building something new. When talking about the (re)construction of territorial and tourist identities, the idea is not to redo them, but to allow the existing form to be reframed. Thus, ter-ritorial and touristic identities in Belém do Pará, where the religious event Círio de Nazaré takes place, are being rethought in a virtual way.

Haesbaert (2007: 156) comments that "territories are not disappearing, but constantly changing places, acquiring another relational sense", this change of place means deterritorializing one space to territorialize another, that is, deterritorializing physical places to territorialize virtual places that contribute to multiterritorialization.

Multiterritoialization is characterized by the connection of territories and the greater ease of spatial movement, facilitating access (both virtually and materially) to the different places where it is anchored (Castells, 1999).

The verbal and visual representations that characterize the virtual places and territories have the physical world as a reference, which is sometimes referenced through generic spatial elements (beaches, cities, lakes), other times by specific places (a certain beach, a city, a specific lake). Virtual territories, the focus of this analysis, are considered multi-user online environments that represent specific places in geographic space. These online environments undergo changes and allow interactions that give them historicity and add an identity characterization (Fragoso et al., 2012).

In the city of Belém do Pará, cited in this chapter due to the Círio de Nazaré event, we can see the gastronomy, the procession, the dances, the typical festivities, which are their own, encompassing native elements that configure themselves as an expression of regional identity (Alves, 2005). When talking about tourist attractiveness, the Círio is observed as the main motivating element, as well as the regional culture. The performance of the Círio attracts devotees, visitors and tourists, and these alter the city's usual territorialities as a result of their travels. The process of touristification of spaces tries to satisfactorily promote cultural exchanges to most social groups, a sense of belonging through devotion, in addition to earning economic gains for the places. Geographical limitations due to social differences will not be an impediment for changes to happen and cyberspace can be implemented. This item will be better presented below.

The Sacred in Cyberspace: The Religious Event Círio de Nazaré

The Círio de Nazaré event is a Catholic procession that has taken place in the capital Belém, in the Brazilian state of Pará, for over 200 years, and is considered one of the greatest symbols of local identity. The history of the event begins with the devotion to the Virgin of Nazareth by the people of Pará, a saint of Portuguese origin worshipped by the Catholic religion.

The story goes that the caboclo took the saint to his house and that the next day she appeared again on the banks of the river. This occurrence would have been repeated several times until it was taken to the Government Palace to be watched by the guards. However, the next day she was again on the banks of the river, with her cloak full of dew and burrs (Muniz, 2021).

The Círio de Nazaré, which was originally just a procession, has become a major event over the years that currently lasts about 15 days. The original path of the procession took the image of the saint, from the Cathedral of Belém to the Basilica Sanctuary, and this path remains in the main procession, but with the expansion of the same procession, many other processions and other activities were incorporated into what we now call the Festa do Círio de Nazaré (Festival of the Taper of Our Lady of Nazareth); one example is the fair, with items of national folklore and typical food stalls (Figueiredo, 2020).

In 2004, Círio de Nazaré received the title of Brazilian Intangible Cultural Heritage from the National Institute of Historic and Artistic Heritage (IPHAN). In 2015, the United Nations Educational, Scientific and Cultural Organization (UNESCO) classified it as Intangible Heritage of Humanity (Muniz, 2021).

The strength of this phenomenon is in the popular participation, that is, the group of *cirianos*, the people who traditionally participate in the Círio, with or without religious affiliation and for various reasons (regional identity, leisure, tourism, among others) that make it a wide-ranging phenomenon of social and cultural importance (Iphan, 2006).

It is in view of this complexity and grandeur that one can understand how the Círio de Nazaré draws 2 million people every year to experience it, and allows us to elect it as a major religious event that generates a tourist flow and deserves further study.

Some activities of the Círio de Nazaré Event, which were held in the physical spaces of the church, go through a virtualization process, from an exclusive Internet domain, and go to cyberspace.

The word cyberspace was evidenced by Gibson in 1984, when dealing with the digital territory, composed of computer networks through which information travelled (Pimentel, 2010), a "virtual interaction space, real in its actions and effects" (Martino, 2015: 11). With the advent of the Internet, the integration of churches to religious practices in cyberspace is now being researched by scholars who are trying to go deeper into the phenomenon (Miklos, 2015). We can see religious institutions appropriating the Internet to promote online religious experiences, which give rise to new temporalities, materialities, spatialities, discourses and rituals (Sbardelotto, 2012).

When talking about religious participation on the Web, Helland (2002) says that there are two forms of access – Religion-online and Online-religion: the first nomenclature refers to the use of the Internet by the individual (faithful) only to search for information related to the church, without interactions with the religious community; the second online-religion nomenclature proposes interaction between individuals (the faithful) following various formats of manifestation available on the Web.

Silva (2014) exemplifies the massive participation of the faithful of the Círio de Nazaré on the Web (the sacred in cyberspace) and says that the main accesses are: on official pages of the basilica, separate pages and specific social networks of the procession (public or private). But he points out that there are other ways to extend the faith without using technology.

Analysing Silva's reports that took place in 2014, we note that the insertion of part of the activities of the Círio de Nazaré Religious Event on the Web took place before the COVID-19 pandemic, (i.e., the organizers of the event), following the dynamics of the event market regarding the use of technologies, had already been innovating, which made it easier to adapt to the restrictions imposed by the World Health Organization (WHO) regarding the non-use of physical spaces for events due to the agglomerations of people, which would enable the transmission of the COVID-19 virus. Based on the idea that the Internet arises in the area of events to support organizers in their pre, trans and post-event management, it was intensified in 2020 and the Círio de Nazaré event, which was hybrid (some face-to-face activities and others online) becomes 100% virtual. Thus, through this transition of territoriality, some activities carried out by the organizers of the event are pointed out, based on technological resources inserted in cyberspace, which contributed to the (re)construction of territorial and touristic identities in Belém do Pará, Brazil.

Círio 360° – Círio Virtual Museum

This event has been taking place since 2018, and the aim has always been to bring the faithful closer to the patron saint. The client, through 3D glasses, sees the saint, the marble, the procession and the entire event virtually, turking 360°. In 2020, this feature was the most used, as the virtual platform was the only way to participate in the event.

The Círio Virtual Museum, part of the Círio 360 project, provides immersion in the events and the possibility of following the complete path of Círio in virtual reality. In addition, it offers access to a huge database of photos, videos and images that explain the origin of devotion through historical images, documentaries, photographic exhibitions and several 360-degree videos of all 13 official processions of the festival (Soares, 2020). The museum has a vast electronic collection of the religious event. People might feel as though they are part of the procession since the works are in 360 degrees (Peres, 2019).

Official page, Facebook and WhatsApp

Since early 2000, the event has its official page, with the domain: www.ciriodenazare.com.br. It contains general information about the event, such as history, event schedule, organizing team/sponsors and supporters, information about the basilica and tourist attractions to be seen (Silva, 2015).

With the establishment of the virtual pilgrimage in Nazaré in 2013, social networks started being analysed to ensure the effective participation of the faithful. According to the number of "likes" that appear on the official Facebook profile, it was discovered that it topped 100,000 (Silva, 2015).

Mediatization via digital social networks (Facebook, Twitter, Blogs, etc.) also give rise to a new way of being religious, a new way of being loyal, and a new method of (re)making Cristo de Nazaré via the Internet. The devout use WhatsApp to create chains and share photos and videos (Magalhães, 2017).

The official app of Crio de Nazaré informs the faithful about the history of Crio de Nazaré, its customs, symbols, prayers and music. Its features include the Rosary prayers and the Book of Pilgrimages, allowing believers to follow the daily liturgy on their smartphones (Magalhães, 2017).

KD a Berlinda

"KD a Berlinda" is an application created in 2013, which has gained many followers. The app follows the *Berlinda*, which is the saint in the 12 pilgrimages existing in Círio de Nossa Senhora de Nazaré. The new version includes a series of features, such as a presence sensor and route (map) tracing. The application is a georeferenced system where the location coordinates of the berlinda are captured by GPS, in real time, allowing the faithful to follow the path and time of each procession. It is available on three platforms: Android, iOS (iPhone and iPad) and Windows Phone (Magalhães, 2017).

The app includes tourist information, Pará food suggestions, religious music in honour of Círio, a procession guide, an interactive mural, the saint's current position, and the entire festival programme (Silva, 2015a).

Due to COVID-19 security protocols, and the prohibition of traditional pilgrimages, the KD a Berlinda application underwent some changes: the removal of the map that showed the location of the image, offering monitoring of the event only by TC Cultura and Rádio Cultura. In addition, a photo gallery was included, and the user can make personalized selfies with the APP mask and post them on their social networks (Flávio, 2020).

It was noted that several changes were made in the Círio de Nazaré event, moving from the physical event to the virtual event using technologies on a large scale, (re)building the territorial and touristic identities of the city of Belém do Pará through the Círio de Nazaré event.

These changes need to be monitored and analysed. Will the new routines abruptly created as a result of the COVID-19 pandemic be maintained? How have these changes positively/negatively impacted our reality and especially the Círio de Nazaré event? It is important to answer these and countless other questions.

Final Considerations

The Círio de Nazaré religious event, held in the city of Belém do Pará, represents an expression of faith, a space of devotion and tourism, is part of the popular festival circuit and is an identity event for the place, drawing thousands of people travelling for reasons religious and mental and spiritual relaxation.

Nowadays, technologies have started creating new forms of social relations and modify territories, which are no longer physical and become virtual and contribute to multiterritorialization. The shift from religious rituals to cyberspace implies a change in the traditional models of rites. It was observed that cyberspace presented itself as a favourable environment for the emergence of new virtual communities, but it does not mean the end of the face-to-face experience. What should be thought about is virtualization and the importance it represents, but in every situation, there are favourable and unfavourable actions.

However, for this study, technology was seen to be inserted in the religious event since the year 2000, but slowly. In 2020, due to COVID-19, and the restrictions imposed by the WHO, the event was seen to move from the in-person to the virtual modality, the emptying of the city, the loss of direct and indirect gains related to religious tourism in favour of the Círio de Nazaré event, the mix between space and cyberspace that modifies the interactions between individuals, reconfigured in major social, cultural and economic changes, which need to be accompanied by accurate data, which will guide the planning and management of this important event.

We conclude that virtual events are innovative experiences, that the cities, in this context of change, reconfigure themselves, trying to adapt to the technological and within the changes inserted in cyberspace that contribute to the (re) construction of territorial and tourist identities in Belém do Pará – Brazil, there is Círio 360°, the Círio Virtual Museum, the app on Google Play, Facebook, WhatsApp and KD a Berlinda. We conclude that virtual events are innovative experiences and that the technologies used in cyberspace allow identities to be reconfigured.

From an interdisciplinary perspective, this chapter presents inputs in the area of tourism, technologies, events (i.e., religious events) in the context of the largest religious event in Brazil. Future studies should move towards developing studies of a quantitative nature, through the administration of questionnaire surveys in order to understand the impact of the influence of technology on the symbolic and affective relationship with a particular place or tourist event (e.g., structural equation models).

References

Alves, I. (2005). A festiva devoção no Círio de Nossa Senhora de Nazaré. *Dossiê Amazônia Brasileira II*, DOI: https://doi.org/10.1590/S0103-40142005000200017

Aragão, I.R. (2014). Reflexões a cerca do turismo cultural – religioso e festa católica no Brasil. *Revista Grifos*, 36/37, 53–67.

Augé, M. (2017). *Não-lugares: uma introdução a uma antropologia da supermodernidade.* São Paulo: Papirus.

Bauman, Z. (2008). *Vida para consumo: a transformação das pessoas em mercadoria.* Rio de Janeiro: Zahar.

Castells, M. (1999). *A sociedade em Rede – A Era da Informação: economia, sociedade e cultura.* São Paulo: Paz e Terra.

Castrogiovanni, A.C. (2004). Existe o espaço Turístico? Construções Teóricas no Campo do Turismo: Anais do II Seminário de Pesquisa em Turismo do Mercosul, pp. 9–31.

Dowson, R. (2018). Towards a definition of Christian MegaEvents in the 21st century. *International Journal of Religious Tourism and Pilgrimage*, 5(3), 1–18.

Dowson, R. (2021). An exploration of the theological tensions in the use of churches for events. *International Journal of Religious Tourism and Pilgrimage*, 9(3), 48–69.

Figueiredo, K. (2020). Festa do Círio de Nazaré. *Info Escola.* Retrieved from: www.infoescola. com/datas-comemorativas/festa-do-cirio-de-nazare/ (accessed 5 September 2021).

Flávio, L. (2020). Prodepa realiza mudanças no "Kd A Berlinda" para o Círio de Nazaré virtual. *Agencia Pará.* Retrieved from: https://agenciapara.com.br/noticia/22575/ (accessed 6 September 2021).

Fragelli, C., Irving, M. De A., and Oliveira, E. (2019). Turismo: fenômeno complexus da contemporaneidade? *Universidade Federal do Rio de Janeiro, Caderno Virtual de Turismo.* Retrieved from: www.redalyc.org/journal/1154/115461709017/html/ (accessed 22 August 2021).

Fragoso, S., Rebs, R.R., and Barth, D.L. (2012). Territorialidades virtuais: dentidade, posse e pertencimento em ambientes multiusuário online. *Campós: Associação Nacional dos Programas de Pós-Graduação em Comunicação.* Retrieved from: http://compos. com.puc-rio.br/media/gt1_suely_fragoso.pdf (accessed 9 September 2021).

Fratucci, A.C. (2014). Turismo e território: relações e complexidades. Caderno Virtual de Turismo. Edição especial: Hospitalidade e políticas públicas em turismo. *Rio de Janeiro*, 14(1), 87–96.

Góis, M.P.F. (2020). Turismo, território e urbanização: uma reanálise do caso do município de Angra dos Reis e da região turística da Costa Verde (RJ). *Geo UERJ, Rio de Janeiro*, 37, e33263, 1–26.

Griffin, K., and Raj, R. (2017). The importance of religious tourism and pilgrimage: Reflecting on definitions, motives and data. *International Journal of Religious Tourism and Pilgrimage*, 5(3), 1–9.

Haesbaert, R. (2004). Dos múltiplos territórios à multiterritorialidade. In Anais do I Seminário Nacional sobre Múltiplas Territorialidades. *Porto Alegre: Programa de Pósgraduação em Geografia da UFRGS*. Retrieved from: www.uff.br/observatoriojovem/sites/default/files/documentos/CONFERENCE_Rogerio_ HAESBAERT.pdf (accessed 27 August 2021).

Haesbaert, R. (2007). *O mito da desterritorialização: do "fim dos territórios" à multiterritorialidade* (3 ed.). Rio de Janeiro: Bertrand Brasil.

Helland, C. (2002). Surfing for Salvation. *Religion*, 32, 293–302.

Iphan (2006). *Círio de Nazaré. – Rio de Janeiro* (pp. 1–105). Retrieved from: http://portal.iphan.gov.br/uploads/publicacao/PatImDos_Cirio_m.pdf (accessed 5 September 2021).

Kocyìgit, M. (2016). The role of religious tourism in creating destination image: The case of Konya Museum. *International Journal of Religious Tourism and Pilgrimage*, 4(7), 21–30.

Lévy, P. (1996). *O que é o virtual?* São Paulo: Editora 34.

Magalhães, T. (2017). Círio de Nazaré na era tecnológica: Maria é mobile e se faz GIF. *Instituto Humanitas Unisinos*. Retrieved from: www.ihu.unisinos.br/78-noticias/572493-cirio-de-nazare-na-era-tecnologica-maria-e-mobile-e-se-faz-gif (accessed 5 September 2021).

Martino, L.M.S. (2015). *Teoria das Mídias Digitais: linguagens, ambientes e redes* (2 ed.). Petrópolis, RJ: Vozes.

Miklos, J. (2015). O sagrado nas redes virtuais: a experiência religiosa na era das conexões entre o midiático e o religioso. *V Congresso Internacional de Comunicação e Cultura – São Paulo*. Retrieved from: https://cisc.org.br/portal/jdownloads/comcult/jorge_miklos.pdf (accessed 2 September 2021).

Ministério da Cultura e Turismo (2015). *Viagens motivadas pela fé*. Retrieved from: www.kultur.gov.br (accessed 13 July 2021).

Muniz, C. (2021). Círio Nazaré: a maior festa religiosa do Brasil. *Toda Matéria*. Retrieved from: www.todamateria.com.br/cirio-nazare/ (accessed 4 September 2021).

Nascimento, A.F. do, and Souza, V.C. De. (2019). O turismo religioso na sociedade líquido-moderna: apropriação da fé pelo trade turístico. *Estudos de Religião*, 33(2), 291–314.

Nyaupane, G.P., Timothy, D.J., and Poudel, S. (2015). Compreendendo os turistas em destinos religiosos: A Perspectiva da Distância Social. *Gestão do Turismo*, 48, 343–353.

Peres, s. (2019). Museu Virtual do Círio sai do Pará e chega a Brasília. *Correio Brasiliense*. Retrieved from: www.correiobraziliense.com.br/app/noticia/cidades/2019/11/06/interna_cidadesdf,804076/museu-virtual-do-cirio-sai-do-para-e-chega-a-brasilia.shtml (accessed 5 September 2021).

Pimentel, M.M.C.B (2010). A territorialidade e a dimensão participativa na ciberdemocracia: o caso do Fórum Social Mundial. *Dissertação de Mestrado em Planejamento Territorial e Desenvolvimento Social da Universidade Católica do Salvador* (pp. 1–149). Retrieved from: http://ri.ucsal.br:8080/jspui/bitstream/123456730/248/1/Dissertacao%20Marcia.pdf (accessed 6 September 2021).

Sá, T. (2014). Lugares e não lugares em Marc Augé. *Tempo social Revista de sociologia da USP*, 26(2), 209–229.

Saquet, M.A. (2009). Por uma abordagem territorial. In M.A. Saquet and E.S. Sposito (Eds.), *Território e Territorialidades: teorias, processos e conflitos* (pp. 73–94). São Paulo: Expressão Popular.

Saraiva, A.L. (2010). Religiosidade popular e festejos religiosos: aspectos da espacialidade de comunidades ribeirinhas de Porto Velho, Rondônia. *Revista Brasileira de História das Religiões*, 7, 147–164.

Sbardelotto, M. (2012). *E o Verbo se fez bit: A comunicação e a experiência religiosas na internet* (1 ed.). Aparecida: Editora Santuário.

Silva, A.N. da. (2014). *Devoção na rede: o ciber-Círio. 9° Interprogramas de Mestrado em Comunicação da Faculdade Cásper Líbero*. Retrieved from: https://casperlibero. edu.br/wp-content/uploads/2014/04/Ariana-Nascimento-da-Silva.pdf (accessed 4 September 2021).

Silva, A.N. da. (2015). A romaria virtual de Nazaré. *Dissertação de Mestrado Apresentada ao Programa de Pós-Graduação em Comunicação da Universidade Paulista*. São Paulo: Universidade Paulista.

Silva, A.N. da. (2015a). Rituais virtuais: os aplicativos do Círio de Nazaré. *Anuário Unesco/ Metodista de Comunicação Regional*, 19(19), 111–119.

Silva, M.L. da., and Tourinho, H.L.Z. (2017). Território, territorialidade e fronteira: o problema dos limites municipais e seus desdobramentos em Belém/PA. Urbe. *Revista Brasileira de Gestão Urbana (Brazilian Journal of Urban Management)*, 9(1), 96–109.

Silveira, E.S. da. (2007). Turismo Religioso no Brasil: uma perspectiva local e global. *Turismo em Análise*, 18(1), 33–51.

Soares, B. (2020). *Círio virtual em 360° aproxima os fiéis da padroeira*. Retrieved from: https://redepara.com.br/Noticia/215808/cirio-virtual-em-360-aproxima-os-fieis-da-padroeira (accessed 5 September 2021).

Sousa, B., and Simões, C. (2010). O comportamento e perfil do consumidor do turismo de nichos? *Polytechnical Studies Review*, VIII(14), 137–146.

Vilhena, J. (2002). Da cidade onde vivemos a uma clínica do território. Lugar e produção de subjetividade. *Pulsional: Revista de Psicanálise*, 15(163), 48–54.

Vilhena, J. de, and Novaes, J. De V. (2018). Lugar e não-lugar no mundo virtual: notas sobre criatividade e territórios de existência na rede. *Tempo psicanal*, 50(2), 143–161.

Part VII
Future Research Directions

18 COVID-19 Effects on Tourism Events, Technology Acceleration and Future Research Directions

Muhammad Khalilur Rahman, Rolee Sifa and Azizul Hassan

Introduction

The novel coronavirus or COVID-19 has brought an inevitable tragedy. Numerous people have acutely suffered from illness, unemployment, financial crisis, breaking health system, deaths of loved ones and poverty. The COVID-19 pandemic has affected peoples' lives emphatically (Rahman et al., 2021a). It is visible to some transformative changes in social interactions, professional dynamics and digital relationships (Rahman et al., 2021b). The travel and tourism sector (Rahman, 2019; Rahman and Zailani, 2017) is not any different from this unavoidable situation and it led to an immense loss (Rana et al., 2020; Rahman et al., 2021c). Hundreds of countries suspended international flights overnight. Therefore, many businesses were closed on temporary basis which tends to lead to permanent closures. The businesses operating activities to a short extent merely survived without making revenues. Consequently, employees from travel and tourism events lost 1 million jobs every day.

The tourism industry had been hit hardest by coronavirus impacts and it seems to remain in uncertainty due to the coming waves at any time (Rahman et al., 2021d). It is expected by the Organisation for Economic Co-operation and Development (OECD) (2020a) that tourism activities have fallen by nearly 80% in 2020. Economics of some countries that significantly depend on international, business and events tourism are affected intensely in mitigating tourism and travel operations in many coastal, regional and rural areas. Nevertheless, the vaccination program raises the hopes for the recovery of the financial crisis (OECD, 2021), yet it has remained a challenging issue for the survival of many businesses. Hence, many countries have reopened internal tourism by opening domestic flights to minimize the loss for some particular destinations. Although it is a matter of time and possible that international tourism will restart again, some evaluation, evidence-based solutions and global cooperation are required in order to lift travel restrictions safely. It is also important for governments to support and collaborate with the industry for the survival of businesses as well as the tourism ecosystem, so that financial loss can be minimized.

Altering and bringing new policies might enable the tourism economy (Rahman et al., 2021a) to survive with the continuance of COVID-19 in the

DOI: 10.4324/9781003271147-26

short to medium period, taking some initiatives imperatively from the government level will ease the detrimental effects of the pandemic situation that will provide compatibility to the businesses standing alone in the market (Rahman et al., 2021b). Moreover, the coordination of government alongside the private sectors is also important. The crisis assists in creating opportunities to have contingency plans for the tourism industry in future. The taken steps, measures and actions will be shaped in future tourism. Hence, the government ought to consider the longer-term implications of the crisis through capitalizing on the digital realm, providing support to low-carbon policies and promoting structural and functional transformation which is required to build a more resilient, stronger and more sustainable tourism economy globally.

The innovation and application of digital technologies are substantially changing people's lifestyles in their work, education, travel, daily activities and business and this is reshaping and transforming tourism (Rahman and Hassan, 2021a; Rahman et al., 2021d). The scope of technology penetration may differ from country to country and sector to sector. As a consequence, the barriers and opportunities that has created unbalanced tech-driven practices in between traditional small and micro businesses and hi-tech implicated tourism businesses that exacerbated the situation by creating gaps. Digital marketing and e-commerce have grabbed attention in building brands, reaching new markets and engaging customers with the products or services (Rahman and Hassan, 2021b). Although existing technologies can create awareness, facilitate financial transactions and promote connectivity among the business operation, these could be less effective to the competitive global marketplace without embracing innovation and enhancing productivity.

Productivity enhances through technologies such as data analytics, revenue management software and cloud computing that have received a low uptake in tourism in general, whereas innovation of new technologies including geotagging and augmented reality has generated more products, services and experiences through novel ways through customizing and delivering the needs of consumers (OECD, 2018). This digital transformation is thus pushing and assisting tourism operating systems to a new era that could boom in the business. Therefore, digitalization has intended to implement in all areas of the tourism industry including operations and value chains. Enabling and facilitating digital technologies in the tourism industry is the key evaluation in adopting new policies to face the challenges of the COVID-19 crisis. This study has aimed to explore the future research pathways within technology acceleration, the COVID-19 epidemic and its effect negatively on tourism events.

COVID-19, Tourism and Tourism Events

COVID-19 background

The World Health Organisation (WHO) declared the novel coronavirus a global pandemic in March 2020, following its identification in Wuhan, China in December 2019 (Shahi et al., 2021). As of 1 September 2020, the number of

COVID-19 cases was over 25 million worldwide, with the fatalities surpassing 850,000 globally (Clinical Trials Arena, 2021). While the COVID-19 case was escalated to a pandemic situation, numerous countries enforced lockdowns and stay-at-home policies as well as border closures, affecting 3.9 billion people globally which is almost half of the world's population (OECD, 2021). In addition, some countries including India, Italy, UK and New Zealand imposed mandatory confinements and some other countries consisting of Mexico, Germany and Canada urged their citizens to stay at home with or without enforcement measures. Many countries such as Chile, Panama, Kenya and Egypt ruled out curfews, whilst residents of places such as Helsinki, Kinshasa and Riyadh were banned from travel to any other localities. The potential of spreading COVID-19 slowed due to lockdown approaches which acted as an effective mechanism to control the highly contagious coronavirus.

COVID-19 effects on tourism events

The global outbreak of COVID-19 has deeply affected the tourism and tourism events. COVID-19 halted the staging of travel and tourism events activity around the world, with events either cancelled, postponed, or rescheduled as virtual or hybrid events. The travel and tourism events sector advises there is a high likelihood of a slow recovery given confidence issues, and the long lead time for events. The COVID-19 pandemic causes the cancellation or postponement of many tourism events in the period pandemic crisis. The community, sporting, cultural and tourism events are a key driver for generating the revenue of any country. The travel and tourism events can help drive a country's national and international positioning, tourism and business linkage. Tourism events provide international and interstate broadcast exposure and a high level of spending from event organizers and participants. Community, sporting and cultural events are crucial for community connection and wellbeing and provide jobs for the tourism event supply chain across the globe. Prior to the COVID-19 pandemic and lockdowns, many travel and tourism events organizations generated economic benefits, attracted millions of tourists and contributed many local and international jobs as well as contributed significantly to the country's economy.

Although lockdowns have helped to slow down the spread of COVID-19, the measures have halted some travel and tourism events. As of April 2020, the number of people who live in countries with travel bans is 7.1 billion, 39% of the global population are living in countries which closed their borders for non-residents and non-citizens, including India and China, among others (Pew Research Center, 2020). As a consequence, as of May 2020, three countries out of four countries and territories globally suspended travel in between different destinations. It has also been estimated that the number of tourist arrivals might drop by 60–80% in 2028 (Aburumman, 2020). Additionally, WTTC had also claimed that the pandemic situation has impacted 121 million jobs in the international tourism sector, with a massive loss in global GDP of around US$3.4

trillion (Wyman, 2020). These remarkable reductions of revenues in the organization were intended to cut off employees for poor financial situations. WTTC had estimated that the loss of jobs was 1 million per minute every day during the peak of the crisis. In the tourism sector, 80% of businesses that are Small and Medium-Sized Enterprises (SMEs) had still their operational activities (Wyman, 2020). Many larger businesses suffered from financial downturns. According to the report of OECD (2020b), many companies reported bankruptcy; for example, around 513 restaurant businesses, 297 transportation providers and 117 aviation companies went through such a bankruptcy situation. Wyman (2020) reported that it has been estimated that the aviation industry faced a net loss of $84.3 billion in 2020; while tour operators anticipated 50% of tours as a result 50% of revenue declined in comparison with the year 2019.

Global travel and tourism events had been bitten hard by the COVID-19 pandemic crisis which left unprecedented effects on employment and businesses. Travel and tourism events are one of the top affected sectors whereas coronavirus nearly halts all types of tourism and tourism events activities all over the world due to border closures and strict lockdown. This sector is being one of the last to recover until withdrawing travel restrictions and overcoming the global recession (OECD, 2020b). As a result, travel and tourism events is in significant trouble along with others sectors that are closely related and dependent on the tourism and tourism events sectors. Once the travel restriction has been eased, it remains a challenge to cover up the tremendous losses over an uncertain time. The rolling out of vaccines might help sometimes to overcome a situation; however, it is plausible that the pandemic situation will be repeated due to the variant nature of coronavirus. This will lower travellers' confidence and damage tourism events and existing businesses.

COVID-19, Technology Acceleration and Tourism Events

COVID-19 pandemic and digitalization

According to United Nations Conference on Trade and Development (UNCTAD) (2021), COVID-19 is acting as the accelerated catalyst in thriving for innovation and the integration of new technologies in the travel and tourism events (UNCTAD, 2021). The COVID-19 pandemic is exposing a large divide between high-income/upper-middle-income countries and poorer countries in digital usage and the availability of technology solutions to support the pandemic response. Ruling out stay-at-home almost in all the countries, it intends to adopt and consume digital media to an extent whereby consumers are expecting to experience contactless technologies consisting of biometrics, one of the basic prerequisites for seamless and safe travel experiences. Therefore, the demand for cybersecurity has also become more important as a part of remote work or work from home where it needs to be ensured the security of identities of digitalized versions (UNCTAD, 2021). Digitalization offers numerous opportunities within precautions to assure employees from the local community working simultaneously. It has a positive overview of adopting digitalization during this crisis.

Digital trends in the tourism and tourism events

Technology acceleration is the process of conversing the information in digital form, which is driven by data management transformation systems to the social and economic systems and lives (OECD, 2020a). The improvement of technologies setting the trends to adopt digitalization to computing data and other information of the travel and tourism events that increase the social and economic connectivity to unfold through globalization. Technology acceleration in tourism and tourism events is intended to boost innovation, generating economic and environmental efficiency and increasing tourism company's outcomes through booming productivity consisting of heightened globalized tourism events (OECD, 2020b).

The business models from any tourism and tourism events adopting new technologies to operate their business in any range within the effective participation in the global valued ecosystem, and introducing of new data-driven approaches, will bring more productivity to shape social and economic wellbeing in the future (Tong and Gong, 2020). Technology acceleration provides supports to the business in transforming data to digital forms through penetrating technologies that make business models and practices, and value ecosystems. For example, the sharing economy including rideshares and accommodation shares has evolved over the past 10 years due to the innovation of new technologies and business model innovation that also has created new worth from the potential travel and tourism events sectors. OECD (2020a) reported that the value of the rideshare sector was at US$61 billion in 2019 whilst the value of accommodation share is estimated to be US$40 billion by the year 2022.

Technology acceleration and opportunity of tourism events

Technology acceleration is embracing the opportunities for tourism and tourism events organizations, Small and Medium-sized Enterprises (SMEs) to bring fortune and develop tourism and tourism events, products and services, innovating business models and procedures, reviewing and upgrading their place at the international level where integrating into digital ecosystems and tourism value chains (UNCTAD, 2021). In addition, technology acceleration has offered potential advantages to SMEs which can be beneficial through the efficient usage of digital form activities, resources focusing on strategic tasks, enabling the development of new business models, and entering new markets and spreading travel and tourism events businesses internationally. SMEs that do not invest in making business digitalized will be left behind in near future in the competitive market of travel and tourism events (OECD, 2018). Therefore, destinations, wider tourism and tourism events, and business are required to immerse in new technologies to survive in the competitive markets. Furthermore, policymakers should act proactively upon shifting towards new technologies for the revolution from traditional systems to the digital system for maintaining potential productivity, innovation and potential value creation for the travel and tourism events.

COVID-19 Effects and Future Research Directions for Tourism and Tourism Events

Technology-enabled tourism events

The progression of technology has had a deep impact on travel and tourism events. These innovative technologies range from technologies that support, create innovative tourism products, services and capturing tourism events (e.g., artificial intelligence, cloud computing, Internet of Things, augmented/virtual reality) to the tourism and tourism events business management systems (e.g., blockchain, smartphone technologies/cloud computing, advance robotics, automation and data analytics) (Rahman and Hassan, 2021a; OECD, 2018). Cloud technologies, international mobile plans including roaming and Internet access in smartphones helps consumers to make travel plans easier than before throughout the whole travel period by tracking and storing the information which is related to their travel (Rahman and Hassan, 2021a). Additionally, online booking, real-time information and mobile payment are also helping customers in ensuring a convenient journey. Cloud technologies allow tourism events to operate business from anywhere in the world within the connectivity of the Internet.

In this digital age, businesses and consumers are alike in generating everlasting new data. The tourism business capacity and tourism events is driven to new data towards more productivity and new business models. Data analytics provides the estimation of customers' preferences and consumers' purchasing behaviour as well as tourism and tourism events activities (Rahman and Hassan, 2021a). A database management system is also used in revenue and implying dynamic pricing. SMEs' employees evolve in educating themselves by developing skills, business policies and procedures, driven to ecosystems and data sharing as per the government rules (OECD, 2018). Different forms of voice-over technology, artificial intelligence and chatbots are enabling customers to undertake Internet browsing, digital check-in, accessing concierge services and voice assistant. Moreover, the technologies offer on-demand, customized, personalized services with filter options on the webpage that facilitate hassle-free travel and tourism events.

Internet of Things (IoT) can also facilitate exchanging data from different devices through the Internet and this supports tourists to travel to tourism events in different cities without their physical object or devices (Javaid and Khan, 2021). The embedded data, software and sensors are used to access information for the marketing purpose and managing tourism events and this leads to create a positive experience for visitors, increase operational and resource efficiencies as well as reducing environmental impacts. Singh et al. (2020) state that the virtual objects in the real world could be shown through augmented reality systems. The usage of technology has replaced paper-based marketing and print-based advertising materials, brochures, travel and tourism events assistant guides and map directions. Smart contracts, based on the blockchain, could be administered to the supply chain system. Therefore, the perpetual usage of user-friendly apps for tourism and tourism events has enhanced all sizes of tourism businesses through real-time data consisting of end-to-end user transparency.

Significance of domestic tourism and tourism events

Domestic tourism and tourism events in the COVID-19 pandemic crisis will contribute to boost the sustainability for many tourism businesses, and destinations (Nunkoo et al., 2021; Rahman et al., 2021b), and it is considered as a key driver to heal financial crises in the short to the medium timeframe (Canh and Thanh, 2020). There have been some initiatives in the domestic tourism sector since the mid-year for the reason of shifting and minimizing the effects of coronavirus in the restriction of international travel and tourism events restrictions. Nonetheless, this has been considered as an obstacle for many countries facing another wave of coronavirus, and domestic tourism is estimated to end the year down sharply on pre-COVID levels. For instance, the United Kingdom and Spain had predicted a decrease by 45–50% in 2020, in domestic travel (OECD, 2020a). Moreover, it is not always beneficial for many countries' tourism events and businesses by altering consumer demands and behaviours, easing the restriction through the movements. This is having some concrete social and economic consequences for many workers, travel destinations, and businesses and the wider financial situation. Tourism and tourism events supports in generating foreign exchange, jobs, and business that is driven to innovation, regional development and underpins local communities. As previously mentioned, before the pandemic the tourism sectors directly contributed to GDP, employment and service exports followed by 4.4%, 6.9% and 21.5% according to OECD (2020b) report. However, this contribution is much higher in several countries whereas tourism and tourism events are the key driving factor of their economic activities. According to the World Health Organization, in 2020 the percentage of contribution to GDP in France, Greece, Iceland, Mexico, Portugal and Spain was 7.4%, 6.8%, 8.6%, 8.7%, 8.0% and 11.8% respectively (OECD, 2020a). It is also remarkable for the indirect impacts of tourism and tourism events to exaggerate the size of economic boosting for local or national economies.

Tourism and tourism events policy implications

COVID-19 has brought an unacceptable loss to the tourism and tourism events as well as people's livelihoods. We should consider this huge shock with the full consequences as the pandemic situation continues (Wyman, 2020). Therefore, policymakers are urged to learn from the crisis for establishing stronger and more resilient strategies for the future of tourism and tourism events. This crisis has brought collaborative efforts from the government level to respond in an action for supporting the recovery of the tourism and tourism events, through delivering accessible support and offers that may compensate the vulnerability of tourism events, workers and tourists at the earliest. Therefore, all the governmental and non-governmental sectors are prepared to react and adapt policies and procedures as quickly and efficiently as possible. This requires crisis-responding mechanisms and assessing the risks through the evaluation at any level, either national or international.

It is a mandate to evolve in co-operating and supporting the activation of tourism and tourism events. Many countries need to come to a point where they can work together as a team for the benefits of the global tourism events as to be taken initiatives for travellers and business holders from different countries (OECD, 2021). They need to develop some adaptive strategies beyond the borders, ensuring the safety of the travellers that stimulate demand and accelerate the recovery of tourism and tourism events. These systems need more attention to cope with any future shocks.

Taking essential actions and providing suitable policies will limit the uncertainty of travel and tourism events and it will offer a crucial recovery as well. The overall view of tourism events remains extraordinarily unsure which damages business and travel confidence. The precise communication, well-designed information policy and the clarification of epidemiological criteria will be remarkably important, requiring some changes in maintaining travel restrictions and containment measures to reduce virus outbreaks and keeping sanitization practices. Therefore, research, information gatherings and analytical data will help to decide in an evidence-based manner for the business policymakers to face the crisis. Consistent and reliable indicators are required to deliver solutions for the critical situation that is created for the COVID-19 outbreak. Hence, the effectiveness of programs and initiatives will be implied in the progression of the recovery process promptly. Rahman et al. (2021b) postulated that risk-controlling solutions will ease the travel restrictions and will back the international tourism ecosystem that is based on the facts and scientific evidence. These solutions should contain the feasibility in the application of functional services that are provided sufficiently and reliably.

The COVID-19 pandemic has been exposed once again, whereas the situation gets prone to the vulnerability of the tourism events (Rahman et al., 2021b). There is an emerging demand to strengthen and diversify the resilience of the tourism economy for handling future shocks, addressing long-standing structural weaknesses and shifting to digitalized technologies will play a vital role in the sustainability of tourism and tourism events development. Tourism and tourism events policy needs to work proactively and to become more flexible to the faster adaption in changes for a long term policy. Crisis management needs to be focused on an area precisely. The longer the crisis continues, the financial implication will arise through losing jobs and making a loss in businesses (OECD, 2020a). Therefore, the economy will break down and it will get harder to recover and rebuild. However, it will make the economic situation more challenging and that leads to creating more opportunities for the innovation of ideas, adopting new ideas, creating niches/markets for the new form of tourism and tourism events business that will open up new business destinations as well as making the business resilient and sustainable to the development of tourism models.

This is also an opportunity for penetration of new technologies, implementing less carbon emission, green recovery strategies and shifting to new business standards for balancing environmental and socio-economic impacts of tourism and tourism events. Policymakers should work on leveraging the opportunities to intensify the sustainability without depletion of ecological balance within fairer and stronger business policies. Therefore, this effort will help in combating the crisis assisting in the development of resilient and sustainable models of tourism events.

The service digitalization is expected to continue booming, consisting of contactless payments and other services, penetration of automation systems, virtual experiences and fact-based variants to gather real-time information (OECD, 2020b). Contingent planning has become popular in demanding the travellers' decision-making process. Acquiring information from different sources such as business websites should contain their promotions and updates in regards to accessing information to their clients and customers. The authenticity and transparency of information will build the trust of travellers towards the tourism and tourism events organization. Moreover, this approach will make the tourism and tourism events business identity in conveying uniqueness.

Conclusion

The global pandemic COVID-19 leads the disruptive and detrimental impacts on global health, social and economic emergency. As of December 2020, the World Tourism Organization reported the confidence index in tourism and tourism events operation is at the lowest level due to the difficulty of controlling coronavirus, travel restrictions and low consumer confidence. Consumers' recreational activities and other travel-related activities are highly dependent on tourism and tourism events operations. Consequently, the COVID-19 pandemic has impacted the international tourism market and tourism events through manifesting the perception of tourists on some health and psychological risks due to the uncertainty of the pandemic situation. Governments have mitigated the spreading virus and reduced social contact by minimizing tourism that has affected people's leisure and travel activities by imposing social distancing. The tourism and tourism events providers through redesigned products and services and innovative travel experiences via technological acceleration also means they use it to follow the restricted policies. The continuous spreading of the COVID-19 pandemic has created a crisis in which tourism and tourism events operators have had to rethink the business operational strategies and update the contingent social systems including cooperation, socio-economic responsibilities and fairness that are provided to the community as well as a society via innovative capabilities and strategical adaptabilities.

References

Aburumman, A.A. (2020). COVID-19 impact and survival strategy in business tourism market: The example of the UAE MICE industry. *Humanities and Social Sciences Communications*, 7(1), 1–11.

Canh, N.P., and Thanh, S.D. (2020). Domestic tourism spending and economic vulnerability. *Annals of Tourism Research*, 85, 1–12.

Clinical Trials Arena (2021). *Coronavirus: A timeline of how the deadly COVID-19 outbreak is evolving*. Retrieved from: www.clinicaltrialsarena.com/news/coronavirus-timeline/ (accessed 5 September 2021).

Javaid, M., and Khan, I.H. (2021). Internet of Things (IoT) enabled healthcare helps to take the challenges of COVID-19 pandemic. *Journal of Oral Biology and Craniofacial Research*, 11(2), 209–214.

Nunkoo, R., Daronkola, H.K., and Gholipour, H.F. (2021). Does domestic tourism influence COVID-19 cases and deaths? *Current Issues in Tourism*, DOI: 10.1080/ 13683500.2021.1960283

OECD (2018). Fostering greater SME participation in a globally integrated economy. OECD Ministerial Conference on SMEs. 23 February 2018. Mexico City, www.oecd. org/cfe/smes/ministerial/documents/2018-SME-Ministerial-Conference-Plenary-Session-3.pdf (accessed 25 August 2021).

OECD (2020a). *Rebuilding tourism for the future: COVID-19 policy responses and recovery*. Retrieved from: www.oecd.org/coronavirus/policy-responses/rebuilding-tourism-for-the-future-covid-19-policy-responses-and-recovery-bced9859/ (accessed 4 September 2021).

OECD (2020b). *Preparing tourism businesses for the digital future*. Retrieved from: www. oecd-ilibrary.org/sites/f528d444-en/index.html?itemId=/content/component/ f528d444-en#chapter-d1e8470 (accessed 8 September 2021).

OECD (2021). *The territorial impact of COVID-19: Managing the crisis and recovery across levels of government*. Retrieved from: www.oecd.org/coronavirus/policy-responses/ the-territorial-impact-of-covid-19-managing-the-crisis-and-recovery-across-levels-of-government-a2c6abaf/ (accessed 6 September 2021).

Pew Research Center (2020). *More than nine-in-ten people worldwide live in countries with travel restrictions amid COVID-19*. Retrieved from: www.pewresearch.org/fact-tank/ 2020/04/01/more-than-nine-in-ten-people-worldwide-live-in-countries-with-travel-restrictions-amid-covid-19/ (accessed 5 September 2021).

Rahman, M.K. (2019). Medical tourism: tourists' perceived services and satisfaction lessons from Malaysian hospitals. *Tourism Review*, 4(3), 739–758.

Rahman, M.K., Masud, M.M., Akhtar, R., and Hossain, M.M. (2021a). Impact of community participation on sustainable development of marine protected areas: Assessment of ecotourism development. *International Journal of Tourism Research*, 1–11. https:// doi.org/10.1002/jtr.2480

Rahman M.K., Islam, M., and Jalil, M.A. (2021b). Corporate Social Responsibility and COVID-19 Crisis. In I. Codreanu, M. A. Hossein and H. S. Min (eds.), *Recent Research and Innovation: An Integrated Approach*. New Delhi: DHA Foundation, Bharti Publications, pp. 167–182.

Rahman, M.K., Gazi, M.A.I., Bhuiyan, M.A., and Rahaman, M.A. (2021c). Effect of Covid-19 pandemic on tourist travel risk and management perceptions. *PloS One*, 16(9), 1–18.

Rahman, M.K., Rana, M.S., and Hassan, A. (2021d). Development and Investment in Core Niche Tourism Products and Services in Bangladesh. In A. Hassan (ed.), *Tourism in Bangladesh: Investment and Development Perspectives*. Singapore: Springer.

Rahman, M.K., and Zailani, S. (2017). The effectiveness and outcomes of the Muslim-friendly medical tourism supply chain. *Journal of Islamic Marketing*, 8(4), 732–752.

Rahman, M.K., and Hassan, A. (2021a). Tourists Experience and Technology Application in Bangladesh. In A. Hassan (Ed.), *Technology Application in the Tourism and Hospitality Industry of Bangladesh*. Singapore: Springer.

Rahman, M.K., and Hassan, A. (2021b). Tourism as an Element for Economic Growth in Bangladesh: Investment Analysis for Product and Service Development. In A. Hassan (Ed.), *Tourism in Bangladesh: Investment and Development Perspectives*. Singapore: Springer.

Rana, M.S., Rahman, M.K., Islam, M.F., and Hassan, A. (2020). Globalization Effects on Tourism Marketing in Bangladesh. In A. Hassan (Ed.), *Tourism Marketing in Bangladesh: An Introduction*. Singapore: Routledge, pp. 157–171.

Shahi, G. K., Dirkson, A., and Majchrzak, T.A. (2021). An exploratory study of COVID-19 misinformation on Twitter. *Online Social Networks and Media*, 22, 100104.

Singh, R.P., Javaid, M., Haleem, A., and Suman, R. (2020). Internet of things (IoT) applications to fight against COVID-19 pandemic. *Diabetes and Metabolic Syndrome: Clinical Research and Reviews*, 14(4), 521–524.

Tong, A., and Gong, R., (2020). *The impact of COVID-19 on SME digitalisation in Malaysia*. Retrieved from: https://blogs.lse.ac.uk/seac/2020/10/20/the-impact-of-covid-19-on-sme-digitalisation-in-malaysia/ (accessed 8 September 2021).

UNCTAD (2021). *Technology and Innovation Report: Catching technological waves Innovation with Equity*. Retrieved from: https://unctad.org/system/files/official-document/tir2020_en.pdf (accessed 2 September 2021).

Wyman, O. (2020). *The future of travel and tourism in the wake of covid-19*. Retrieved from: www.oliverwyman.com/content/dam/oliverwyman/v2/publications/2020/To_Recovery_and_Beyo (accessed 7 September 2021).

Conclusion

Azizul Hassan

Digital innovations through event booking websites, movies, blogs, and photography, technology have helped tourism events expand their reach over the previous few decades. Organizers and attendants planning their next tourism event or making an event wish list may find digital innovation tools and material to be a valuable source of knowledge. While remote or virtual digital innovation has long been a future topic in tourism event circles, the world today, influenced by the COVID-19 pandemic, may now be ready to accept it. Major findings of this book are noteworthy to be mentioned in this regard.

The conceptual framework and theoretical underpinnings discussed in Chapter 1 of this book implies that digital technology-enabled tourism events have become unquestionably a highly sophisticated, diverse phenomenon. This study also provides new insights into the tourism and digital innovation debates by looking at the interplay between events and technology. It also offers potential future directions for digital innovation in the management and evaluation of technology-based tourism events.

According to the findings of Chapter 2, to promote the expansion of the tourist business, various ICT tools and approaches are progressively being employed in almost all sub-sectors. Bangladesh, while being a poor country, is not immune to this tendency. Nowadays, ICT plays a vital role in promoting these nature-based attractions as well as administering, safeguarding, and conserving these places. However, following the advent of the COVID-19 pandemic in March 2020, the government has placed certain harsh limitations on travellers, limiting their travels.

The findings of Chapter 3 illustrate the aspects light show technology (LST) attempted to exhibit, the factors influencing LST's landscaping, and the visitor feelings elicited by LST. The findings and recommendations are applicable to tourism festivals and night tours, placemaking, destination attachment, and other fields. These findings can be used to inform the usage of LST in urban night tourism and festival designs in other Asian nations.

In Chapter 4, results show that the expanding mobility of these phenomena has recently attached the Bengali Diaspora of Indian heritage to their country with enormous passion, uniting and connecting them with their counterparts residing in the state, allowing them to virtually join the carnival. The rest of the

DOI: 10.4324/9781003271147-27

globe will be able to participate in the celebration virtually as well. As a result, it adds to the city's recent surge in festival-centric tourism.

Chapter 5 finds that from the standpoint of utilitarian and hedonic motivation, efforts focused on meeting functional and emotional demands must be carefully considered in order to promote the Kaamatan. This is unavoidable for the event's success, since organizers will not only entice people to attend the virtual event, but also to encourage them to join the real event in the future when borders are opened and travel restrictions are lifted. The collaborative development of likings for this event by addressing needs, according to this chapter, will increase incentive to virtually attend the event and better promote not only the festival but also the destination.

The findings of Chapter 6 show that information and communications technology (ICT) has a significant economic impact, especially in developing nations where it is critical to expand enterprises in a cost-effective manner. Coke Fest (CF) was able to create and grow its fanbase by using the Internet to communicate with them. The food industry has seen a reduction in costs and a rise in income. The Pakistan Super League (PSL), on the other side, has helped the economy by attracting sports tourists to Pakistan, which contributes to the country's total GDP. Coke Fest was Pakistan's first digital event, and it was a success because of an online ticketing system that allowed people to log in and watch the event in real time. Coke Fest was the first in Pakistan to make such a large event a success. Both events have also offered up new possibilities for reaching out to clients via social media and online ticketing platforms on a wide scale. Having a fully digitalized ticketing system can assist both PSL and CF in efficiently managing their event and returns in the event of a pandemic. PSL and CF organizers have been able to improve total revenues, encourage tourism, and helping the economy by collaborating with numerous sponsors.

Chapter 7 highlights how the role of technology application in determining the most important components influencing visitors' desire to return to the festival site, as well as the primary traits that drive visitor behaviour, is shown in the case study of a local cuisine festival in Thailand.

The findings of Chapter 8, which is based on a literature analysis, reveal that, despite the fact that Turkish events are outnumbered, foreign visitors' interest in these festivals is fairly restricted. It is necessary to guarantee that the events are more properly advertised to foreign tourists. Various professional promotion actions should be addressed in order to market the events more effectively. Each event will be unique if the newest technology advancements and smart tourist practices are prioritized in these promotional activities. In Turkey, more festival tourist activities will be realized by utilizing new and clever technology.

Chapter 9 explores that the 16 member nations of the SADC are examined in broad terms, with the most active being examined in further depth in terms of their adoption of modern services such as the Internet and digital tourism. According to the literature, South Africa is the most active nation and looks to have a solid approach to adopting digital tourism. As a result, the majority of the literature on digital tourism is written from a South African viewpoint. There are signs that digital tourism is developing in Southern Africa, and that it will likely continue to rise as access to such services improves.

Chapter 10 finds that using social media to share content, using sharing platforms such as Airbnb and Uber, using payment platforms such as PayPal and Apple Pay, using apps to provide visitor information on nearby offerings, and using virtual reality to provide virtual tours of hotels and hotel rooms are all increasing the appeal of the Australian Open. Despite the fact that Australia is currently struggling to overcome the challenges of COVID-19 in order to recoup revenue losses from the previous year, it is expected that this sporting event will show resilience and continue to succeed in the coming years through the use of technology applications.

Sport activities were effectively planned and disseminated to a wide variety of audiences as a consequence of technical developments, allowing fans to follow the games from their homes throughout the pandemic limitations, according to the findings of Chapter 11. With the 2021 Sabre World Cup in Budapest, Hungary, organizers have made it clear that digital technologies are being used to improve the quality of the consumer experience, but destination brand development appears to have untapped potential.

Chapter 12 finds that the examination of the Note di Fuoco event in Calabria, Southern Italy, devoted to the art of pyrotechnics, revealed how much the use of cutting-edge technology allows for more comprehensive and inclusive event experiences. Hearing and touch are very important, especially when 3D viewers are employed to provide completely immersive experiences. Note di Fuoco is an event that employs technology in a variety of ways, both to enhance the user experience and to assist the event's organizers in better managing the event. In the future, research into new technologies that can be used to create and use events will make them even more memorable.

Chapter 13 explores in the Portuguese context, the event "7 Wonders of Gastronomy" and the digitalization of communication. The findings show that combining food and wine is used in tourism attachment and touristic locations, with consequences for destination management and local development.

Chapter 14 pinpoints that the COVID-19 pandemic has had a significant impact on Spain's social quality of life and people's well-being, resulting in changes in the tourist business. However, the country is medically prepared to welcome visitors, is positioning itself as a destination for open-air micro-personalized events, is ready for amazing hybrid high-tech business travel events like the World Mobile Congress, and is mindful of the importance of sustainability and diversity when planning purposeful events.

Chapter 15 outlined the tourism and hospitality industry's economic and other functions across the world, as well as in the United States. It also demonstrated the pandemic's influence on every level of the industry. Scholars and industry professionals believe that technology is one of the most important methods for the sector to adapt and (possibly) become more successful than it was previously.

Chapter 16 finds that the connection and discussion between companies and their goods with the consumer audience altered during Rock in Rio 2019. It was a time for sharing experiences as well as putting technical advancements to the test. The Internet and technical advances at Rock in Rio have become central to this new situation, helping companies interact with their customers and encouraging more participation in the event. Furthermore, when they incorporated a Management Plan based on the pillars of sustainable development (Social, Economic, and Environmental) and highlighted green marketing, the event's sustainability became a reality.

The findings of Chapter 17 show that the cyberspace of the Taper of Our Lady of Nazareth Religious event is an innovative and interactive religious experience that allows for interaction through technology-mediated behaviours. This chapter covers inputs in the disciplines of tourism, technology and events in the context of the world's largest religious event from an interdisciplinary approach.

Chapter 18 explains the COVID-19 effects on tourism events, technology advancement, and future research directions. This chapter makes numerous proposals for rethinking tourism model development, global value ecosystems, and digital acceleration trends in tourism events, as well as a list of critical policy issues for fostering technology adoption and application in the tourism and tourist events industries.

Tourism events are expected to undergo a fundamental transformation in terms of information and engagement technologies. Such events are one such dramatic manifestation in building a tourist offering, and it is the start of a totally fresh digital tourism innovation path. Many tourism events throughout the world rely on digital innovation to run their operations. Digital innovations are being used by major tourism events to improve their capabilities. Recently, countries have been battling to reduce the economic effect of the extraordinary COVID-19 pandemic. Many governments have implemented travel restrictions in order to limit people's movement and prevent the spread of diseases. Travel restrictions have put the tourist business in jeopardy all around the world. If the scenario continues, governments will find it difficult to open tourism. As a result, tourism digital innovation is a possible option in this circumstance. However, it is critical to determine whether digital innovation can give alternative answers and whether digital innovation can replace traditional tourism. Tourism events require major change in order to have a greater impression on potential tourists, and digital innovation progress is the most promising future for events.

Index

For Product Safety Concerns and Information please contact our EU
representative GPSR@taylorandfrancis.com
Taylor & Francis Verlag GmbH, Kaufingerstraße 24, 80331 München, Germany

* 9 7 8 1 0 3 2 2 2 0 9 7 0 *